Beyond the Amber
Waves of Grain

RURAL STUDIES SERIES
of the
Rural Sociological Society

Series Editor

Lionel J. Beaulieu, *University of Florida*

Editorial Board

Jess C. Gilbert, *University of Wisconsin–Madison*
Richard S. Krannich, *Utah State University*
Linda M. Lobao, *Ohio State University*
Rogelio Saenz, *Texas A&M University*
DeeAnn Wenk, *University of Oklahoma*
Deborah M. Tootle, *USDA ERS ARED*

Rural Studies Series

Beyond the Amber Waves of Grain: An Examination of Social and Economic Restructuring in the Heartland, Paul Lasley, F. Larry Leistritz, Linda M. Lobao, and Katherine Meyer

Community and the Northwestern Logger: Continuities and Changes in the Era of the Spotted Owl, Matthew S. Carroll

Investing in People: The Human Capital Needs of Rural America, Lionel J. Beaulieu and David Mulkey

Against All Odds: Rural Community in the Information Age, John C. Allen and Don A. Dillman

Rural Data, People, and Policy: Information Systems for the 21st Century, edited by James A. Christenson, Richard C. Maurer, and Nancy L. Strang

Economic Adaptation: Alternatives for Nonmetropolitan Areas, edited by David L. Barkley

Persistent Poverty in Rural America, Rural Sociological Society Task Force on Persistent Rural Poverty

Rural Policies for the 1990s, edited by Cornelia B. Flora and James A. Christenson

Research, Realpolitik, and Development in Korea: The State and the Green Revolution, Larry L. Burmeister

Beyond the Amber Waves of Grain

An Examination of Social and Economic Restructuring in the Heartland

Paul Lasley, F. Larry Leistritz,
Linda M. Lobao, and Katherine Meyer

with Freddie L. Barnard, Arlo Biere,
Jackie Fellows, Daryl Hobbs, Bruce Johnson,
Kent D. Olson, William E. Saupe,
Raymond D. Vlasin, and Robert Weagley

Westview Press
BOULDER • SAN FRANCISCO • OXFORD

Rural Studies Series, Sponsored by the Rural Sociological Society

All rights reserved. No part of this publication may be reproduced or transmitted in any form or by any means, electronic or mechanical, including photocopy, recording, or any information storage and retrieval system, without permission in writing from the publisher.

Copyright © 1995 by Westview Press, Inc.

Published in 1995 in the United States of America by Westview Press, Inc., 5500 Central Avenue, Boulder, Colorado 80301-2877, and in the United Kingdom by Westview Press, 12 Hid's Copse Road, Cumnor Hill, Oxford OX2 9JJ

Library of Congress Cataloging-in-Publication Data
Beyond the amber waves of grain : an examination of social and
  economic restructuring in the heartland / by Paul Lasley ... [et
  al.].
    p. cm.—(Rural studies series)
  ISBN 0-8133-8930-5
  1. Middle West—Economic conditions. 2. Middle West—Social
conditions. 3. Household surveys—Middle West. 4. Farms—Middle
West. 5. Agricultural industries—Middle West. 6. Agriculture
—Economic aspects—Middle West. 7. Unemployment—Middle West.
I. Lasley, Paul. II. Series: Rural studies series of the Rural
Sociological Society.
HC107.A14B49   1995
338.S77—dc20   95-1552
                                                                              CIP

Printed and bound in the United States of America

∞ The paper used in this publication meets the requirements
   of the American National Standard for Permanence of Paper
   for Printed Library Materials Z39.48-1984.

10   9   8   7   6   5   4   3   2   1

To our children

Rachel and Sarah Lasley
Lori Jean and Leslie Diane Leistritz
Erick Lobao
Anne and Elizabeth Seidler

Contents

List of Tables and Figures	xi
Foreword	xv
About the Contributors	xvii

1	Farm Restructuring and Crisis in the Heartland: An Introduction, *Linda M. Lobao and Paul Lasley*	1
2	The Agricultures of the Midwest and Their Demographic and Economic Environments, *Daryl Hobbs and Robert Weagley*	29

PART ONE
The Farm Enterprise

	Introduction to Part One, *F. Larry Leistritz*	51
3	Financial Characteristics of Farm Operators, *F. Larry Leistritz and Freddie L. Barnard*	53
4	The Process of Adaptation and the Consequences to the Farm System, *Bruce Johnson and Raymond D. Vlasin*	71
5	Plans for Changing the Farm Business and Needs for Training, *Kent D. Olson and William E. Saupe*	87

PART TWO
The Farm Household

	Introduction to Part Two, *Katherine Meyer*	107
6	The Changing Division of Labor on Family Farms, *Jackie Fellows and Paul Lasley*	109

| 7 | The Impacts of Financial Hardship on Familial Well-Being, *Paul Lasley* | 127 |
| 8 | Perceiving Hardship and Managing Life, *Katherine Meyer* | 143 |

PART THREE
Community and Social Interaction

	Introduction to Part Three, *Paul Lasley*	163
9	Community Change, *Arlo Biere*	167
10	Organizational, Community, and Political Involvement as Responses to Rural Restructuring, *Linda M. Lobao*	183
11	Farm Crisis in the Midwest: Trends and Implications, *F. Larry Leistritz and Katherine Meyer*	207
12	Methodology, *Paul Lasley*	221

Appendix: Survey Questionnaires — 233
Index — 251

Tables and Figures

Tables

2.1	U.S. Population and Change by Region, 1960-1990	37
2.2	Non-metropolitan Economic and Demographic Changes Among Regions, 1980-1990	37
2.3	Number of Farms by Gross Farm Commodity Sales Categories, 1987	41
2.4	Percentage of Total Agricultural Products Sold by Each Gross Sales Category of Farms, 1987, by State	42
2.5	Farm Net Cash Returns by State and Size of Farm, 1987	45
2.6	Commodity Production, 1987	47
3.1	Gross Farm Sales and Net Family Income of Respondents, and Percentage Total Family Income from Farming, Off-farm Employment, and Other Non-farm Sources, 1988	56
3.2	Balance Sheet Information: Assets by Type for Respondents, 1988	60
3.3	Balance Sheet Information: Debts by Type for Respondents, 1988	61
3.4	Balance Sheet Information: Owner Equity and Debt-Asset Ratio	62
3.5	Percentages of Respondents with Debt	62
3.6	Distribution of Farms According to Debt-Asset Ratio for All Farmers in Surveys and for Farmers in Surveys with Gross Farm Sales Less Than and More Than $100,000 per Year	65
3.7	Delinquency Rates and Anticipated Financing Problems	66
4.1	Farm Operators' Report of Risk-Reduction Tactics for 1984-1988, North Central Region and Subregions	74

4.2	Farm Operators' Report of Risk-Reduction Tactics for 1984-1988, by Age of Farm Operator, North Central Region	77
4.3	Farm Operators' Report of Risk-Reduction Tactics for 1984-1988, by Economic Size of Farming Unit, North Central Region	80
4.4	Farm Operators' Report of Risk-Reduction Tactics for 1984-1988, by Reported Changes in Stress of Farm Operators' Spouses, North Central Region	82
4.5	Farm Operators' Report of Risk-Reduction Tactics for 1984-1988, by Net Family Income Level, North Central Region	84
5.1	Percentages of All Sample Farmers in the North Central Region Planning to Make Selected Changes in the Near Future	92
5.2	Percentages of All Sample Farmers in the North Central Region Reporting a Need for Information and Training	95
5.3	Percentages of Sample Farmers Planning to Make Selected Changes in the Near Future, by Operator's Age and Subregion	97
5.4	Percentages of Sample Farmers Reporting a Need for Information and Training, by Operator's Age and Subregion	99
5.5	Percentages of Sample Farmers Planning to Make Selected Changes in the Near Future, by Gross Farm Sales and Subregion	101
5.6	Percentages of Sample Farmers Reporting a Need for Farming Information and Training, by Gross Farm Sales and Subregion	104
6.1	Farm Tasks Performed by Spouse and Change in Time Spent Performing Tasks in Past Five Years	113
6.2	Comparison Between Families with Different Employment Status	115
6.3	Relationship Between Decision-Making Responsibility and Off-farm Employment as Reported by Spouse	118
6.4	Relationship Between Spouse's Farm Tasks and Off-farm Employment as Reported by Spouse	121
6.5	Relationship Between Change in Time Spent on Tasks in Past Five Years and Off-farm Employment as Reported by Spouse	123

7.1	Adjustments Reported as Made Because of Financial Need in Past Five Years	132
7.2	Relationship Between Debt-Asset Ratio and Quality of Life Measures	132
7.3	Relationship Between Income and Quality of Life Measures	134
7.4	Relationship Between Debt-Asset Ratio and Household Adjustments	135
7.5	Relationship Between Household Adjustments and Perceived Quality of Life and Satisfaction	136
7.6	Relationship Between Socioeconomic Characteristics and Household Adjustments, Zero-Order Correlations	138
8.1	Sources of Daily Pressures: Rotated Factor Pattern	150
8.2	Coping Strategies: Quartimax Rotated Factor Pattern	153
9.1	Farm Income by State, in Millions of Dollars	169
9.2	Farm Operators' Responses to Changes in Community Conditions, 1984 through 1988, by County Economic Type	173
9.3	Multiple Comparisons Tests of Farm Operators' Responses to Community Changes by County Economic Type	174
9.4	Farm Operators' Responses to Changes in Community Conditions, 1984 through 1988, by Metropolitan Proximity and County Population	177
9.5	Multiple Comparisons Tests of Farm Operators' Responses to Community Change by County Size and Proximity	179
10.1	Membership in Community and Farm Organizations	189
10.2	Community Work, Political Activism, and Protest	192
10.3	Organizational and Political Participation and Its Correlates (Operators)	195
10.4	Organizational and Political Participation and Its Correlates (Spouses)	196
11.1	Standardized Discriminant Coefficients of Selected Measures on Gross Farm Sales, by Set	214
12.1	Information for Weighting Regional Samples (Operators)	227
12.2	Information for Weighting Regional Samples (Spouses)	228

12.3	Comparison of Personal and Farm Characteristics of Regional and Subregional Samples to U.S. Census of Agriculture	229

Figures

1.1	Dimensions of change during the crisis period	17
2.1	Real interest rates 1975 through 1988	30
2.2	Soybean, wheat, and corn prices 1975 through 1988	31
2.3	Beef, chicken, pork, and milk prices 1975 through 1988	31
2.4	Total exports in billions 1975 through 1988	32
2.5	Average farm land values by state 1975 through 1988	33
2.6	Average farm land values by state 1975 through 1988	34
3.1	North Central Region and Subregions	55
3.2	Sources of family income for respondents with gross farm sales less than $100,000, sales of $100,000 or more, and all respondents, North Central Region, 1988	58
3.3	Degree of financial leverage for respondents with gross farm sales less than $100,000, sales of $100,000 or more, and all respondents, North Central Region, 1988	64

Foreword

Economic restructuring is increasingly documented; however, the human consequences are often overlooked. When they are examined, the focus is on the labor-capital nexus, looking particularly at impacts on individuals who are laid off from the formal sector. The self-employed are an important sector in rural areas, disadvantaged in a globalizing economy that favors the large and the mobile. The small and the rooted, such as the midwestern farm households, are the focus of this study.

Beyond the Amber Waves of Grain is the first systematic account that locates the farm crisis of the 1980s within the larger historical trend of rural restructuring. Throughout most of the post-World War II period, farming has undergone a dramatic shift from being organized in a dispersed system of family farms to increasingly large-scale businesses. Many observers have referred to this transformation as moving from "farming as a way of life" to "farming as a business." It has profoundly affected the social and economic fabric of rural America. Rather than an aberration, the 1980s farm crisis was an acceleration of historical restructuring of farming and rural life.

The marvels of our technologically advanced agriculture have kept yields high. Yet, the transformation of agriculture has not been without substantial costs to rural communities and farm families. Based upon a study of more than 7,000 farm men and women across the 12 North Central states, the region hit hardest by the crisis, the authors focus on changes within farm enterprises, farm households, and rural communities. Through systematic data collection and analysis, we see another side of farming, one that is reflected in many national statistics on out-migration, farm consolidation, abandoned schools, churches, and small businesses.

The North Central Regional Center for Rural Development is pleased to have been a part of this effort. This study was initiated under the direction of my predecessor, Peter Korsching, who facilitated formation of the study team and helped guide it through the early steps. The project received valuable counsel from the Center's Board of

Directors and received financial support through the Center. Organized as a North Central Regional Research Project (NC-184), this was truly an interdisciplinary and interorganizational effort. Each of the land grant universities in the 12 North Central states participated in the study. Social scientists from a variety of disciplines collaborated in this effort to document the farm, familial, and community adaptations to economic hardship.

Major partners in this effort were the agricultural statistics services in each of the states. Duane Skow and the staff at the Iowa Agricultural Statistics Services coordinated the data collection with their counterparts in each of the states. Without their assistance this study would not have been possible. This study demonstrates that it is possible to overcome disciplinary and organizational boundaries that often separate the academic and policy arenas.

As part of the Rural Studies Series sponsored by the Rural Sociological Society, the manuscript has benefited from the diligent surveillance of Forrest Deseran and other, anonymous reviewers. We are grateful to the many farm families who completed the questionnaires and shared their concerns. To the many others who made contributions to this study, we are indebted.

Cornelia Flora
Director, North Central Regional Center
for Rural Development

About the Contributors

Freddie L. Barnard is an associate professor and extension economist at Purdue University and is the director of the Agricultural Banking School held at Purdue University. He also serves as the coordinator for the Financial Criteria and Measures Subcommittee of the Farm Financial Standards Task Force. He holds a Ph.D. in agricultural economics, with a specialty in agricultural finance, from the University of Illinois.

Arlo Biere is a professor of agricultural economics at Kansas State University. Professor Biere's research has been in the areas of natural resource economics, price analysis, local government management and policy, and rural community economics. He also teaches Agricultural Market Structures and Research Methods in Economics.

Jackie Fellows is on the faculty in the Department of Sociology and Social Work, Concordia College, Moorhead, Minnesota. She teaches courses in Introductory Sociology, Social Problems, and Criminal Justice. She is a doctoral candidate completing her Ph.D. dissertation at Iowa State University, Department of Sociology.

Daryl Hobbs is a professor of rural sociology at the University of Missouri and director of the University's Office of Social and Economic Data Analysis. A native of Iowa, Hobbs received his education at Iowa State University. He is past president of the Rural Sociological Society and has been honored by the Society as Distinguished Rural Sociologist.

Bruce Johnson is a professor of agricultural economics at the University of Nebraska-Lincoln. He holds a B.S. and M.S. from the University of Nebraska and a Ph.D. from Michigan State University. His teaching and research interests focus on agricultural finance and taxation, structural change in agriculture, and rural development.

Paul Lasley is a professor of sociology at Iowa State University where he has a joint teaching, research, and extension appointment. His research and teaching program addresses linkages between the agriculture and rural communities with attention to the human dimensions in the food system. He received his education at the University of Missouri.

F. Larry Leistritz is a professor of agricultural economics at North Dakota State University. He has directed numerous research projects addressing various aspects of agricultural and resource economics. He is the author or editor of ten books, including *Rural Economic Development, 1975-1993* (Greenwood 1994) and *The Socioeconomic Impact of Resource Development: Methods for Assessment* (Westview 1981).

Linda M. Lobao is an associate professor of rural sociology at Ohio State University. Her work focuses on economic change and its social and political consequences. Presently, she is collaborating with Katherine Meyer on a study of women's and men's social and political responses to farm and other economic changes in Midwestern communities.

Katherine Meyer is an associate professor of sociology at Ohio State University. Her work centers on social change, the process through which it develops and operates, and its impact on individuals. Her research centers on structural antecedents and structural and social psychological outcomes of change.

Kent D. Olson is an associate professor in the Department of Agricultural and Applied Economics, University of Minnesota, St. Paul. His duties include teaching and research in farm management and the structure of agriculture.

William E. Saupe is a professor in the Department of Agricultural Economics at the University of Wisconsin-Madison/Extension. He received his Ph.D. in agricultural economics from Iowa State University in 1965 and has since served at the University of Wisconsin-Madison.

Raymond D. Vlasin is a professor in the Department of Resource Development, Michigan State University. At the time research was being conducted for this book, Dr. Vlasin was engaged in research on community economic development, supported by the Michigan Agricultural Experiment Station, and served as a specialist in economic development for MSU Extension. Currently, Dr. Vlasin is coordinator of industrial extension for MSU Extension and is helping design and implement the Michigan Industrial Extension Partnership, involving educational, governmental, and private-sector members.

Robert Weagley, Ph.D., CFP, is an associate professor of consumer and family economics at the University of Missouri-Columbia. His research focuses on the areas of household investment and insurance management as well as the economics of human capital development and its allocation.

1

Farm Restructuring and Crisis in the Heartland: An Introduction

Linda M. Lobao and Paul Lasley

A dramatic restructuring of the farm sector has been underway since World War II. This restructuring is evident to the most casual observer throughout many parts of the rural Midwest. Abandoned farmsteads, deserted rural schools and churches, and boarded-up businesses tell the story of changes in farming and its effects upon the rural culture. Whenever hard times come to farming, rural institutions are also affected. Statistics tell a similar, if not equally compelling, story: from 1940 to 1990, the number of farms was reduced by two-thirds and the farm population declined from nearly one-fourth of all Americans to about 2 percent.

Restructuring of farming also has brought fundamental changes in the organization and control of production. Many analysts argue that a dualistic farming system has emerged in the postwar period (Goss, Rodefeld, and Buttel 1980; Krause 1987). That is, an increase has occurred in the relative number of small farms operated part-time and of large farms, particularly those with a half-million dollars or more of annual gross sales (Stam et al. 1991). Moderate-sized "family farms" that use little hired labor and whose households depend mainly on farm income for their livelihood have been edged increasingly out of this system.[1] These farms historically formed the backbone of the traditional rural community.

The declining farm population and family farm and the passing of agrarian community life have long concerned some farm organizations, scholars, and policy makers. Until recently, however, most persons either ignored these changes or regarded them as inevitable trade-offs in creating a modern, highly productive, capital-intensive agricultural system that would serve the needs of domestic consumers as well as international markets. The postwar farm economy was viewed as stable and even as buoyant for a period during the 1970s. An array of programs and policies such as the Farm Credit System, commodity and acreage set-aside programs, and Farm Credit Acts were in place to safeguard farming, protect farm income, and reduce risk. Many of these originated in New Deal legislation, the direct result of the lessons learned from the Great Depression.

The financial "crisis" of the 1980s rocked public complacency about farm restructuring. Americans were confronted with the fact that bucolic notions of farm life did not match the actual hardships and that the Jeffersonian ideal of family farming was shattered. Policy makers discovered that long-standing policies and programs were insufficient and that a massive government bailout of the farm sector would be necessary. Academic and other researchers became aware of their limitations in anticipating the crisis and assessing its magnitude (Harl 1990). Some observers raised more fundamental issues about the trajectory of farm change in the postwar era, including its environmental and social sustainability. Traditional farm organizations were forced to reconsider their platforms and to compete for public attention with newly emerging grass-roots groups. For farmers, the crisis called into question a valued way of life and a possible career path for their children. At the extreme, it meant the loss of household savings, the violation of intergenerational trust whereby farmland passed down through generations, and sometimes the loss of human life.

This book, the story of those who survived in farming during the 1980s, is the first systematic account portraying how the crisis period shaped the lives and enterprises of farm people in the grain-producing heartland of the Midwest. It is based on a study of more than 7,000 farm men and women in the 12 North Central states, the region hit hardest by the crisis. We focus on the changes experienced by farm people in three arenas of rural life: the farm enterprise, household, and community. This focus encompasses the major dimensions along which change was said to occur during the crisis. Empirical findings from the chapters that follow are based on a single data set collected in 1989 from farm operators and spouses. The results provide a comprehensive view of farm restructuring and its consequences and enable us to address a number of questions that have remained generally unanswered about the crisis.

The study differs from most others conducted during this period in that results are broadly generalizable and detailed and are based on observations from both women and men. Most previous studies are limited to a single state or, in the case of national sources such as the Census of Agriculture, provide highly aggregated data which do not permit direct, detailed analyses at enterprise and household level. These studies typically collect information solely from the operator, diminishing the role of farm women who are less likely to be defined, either by themselves or others, as operators. Prior research tends to assess empirically a limited range of crisis-related consequences, with primary focus on economic impacts such as financial stability and enterprise adjustments. In contrast, we deal with an array of issues raised by analysts of the crisis, from the extent of deterioration in objective financial well-being to subjective perceptions about the quality of personal and community life and the behavioral responses aimed at coping with hardship conditions.

Changes in the farm sector have been examined mainly by rural social scientists, including agricultural economists, home economists, and sociologists. Typically, researchers in each discipline work independently, employ distinct methodologies and theoretical frameworks, and focus on different research questions. The farm crisis, however, spilled over into many areas of rural life: the farm as a production unit, the household, and the rural community. Its effects could not be captured by purely disciplinary approaches. If sociologists were to study the outcomes of the crisis for the social well-being of women and men, they had to learn more about the structure of agricultural production. If agricultural economists were to map out adjustments in farm management, they had to be concerned with household composition, the gender division of labor, and decision making. Both economists and sociologists came to recognize that survival in farming would be a function not only of objective farm and household attributes but also of farm people's perceptions, feelings, and collective responses to agricultural change. Thus, the present study is the result of interdisciplinary efforts undertaken by a team of sociologists and economists. As such, it highlights points of focus and research traditions of both disciplines and offers a comprehensive view of the effects of farm restructuring.

The significance of our topic extends beyond the farm gate. First, changes in farming have important implications for related industries and for areas still dependent on farming. Although farmers represent only about 2 percent of the population, farming is the cornerstone of our nation's largest industry, agriculture, which employs nearly one American out of four. Moreover, about one-sixth of U.S. counties depend upon farming for a sizable proportion of local earnings (Carlin 1990: 5). These counties are located mainly in the plains and western corn belt. The ripple effects created by the 1980s crisis jeopardized the non-farm economies and fiscal bases of rural communities located in these regions.

Second, the farm crisis, coupled with downturns in other traditionally rural industries such as mining, oil, timber, and non-durable manufacturing led to a general rural economic recession that continues into the present. The inequality between urban and rural areas which seemed to be diminishing a decade ago, began to widen in the 1980s as evidenced by the increase in poverty, unemployment, and the "working poor" among the rural population (O'Hare 1988). The flooding of large parts of the rural Midwest in the summer of 1993 has further limited the possibility of economic recovery. Focusing on the farm sector, a segment of the rural economy which will continue to experience economic pressure, is important for an understanding of the problems and policy solutions facing many rural areas.

Finally, this study should be viewed as contributing further evidence to the wider body of research on economic restructuring. Most studies on economic restructuring have focused on non-farm industry, paid employment, and urban areas. A relatively large literature exists on the consequences of plant closings, layoffs, and other forms of unemployment. Less is known, however, about the restructuring of self-employment, particularly family enterprises such as farming. Farming often involves unpaid family labor and the use of multiple livelihood strategies that combine formal-sector employment with self-employment such as contracting out with other farmers, as well as home-based businesses such as machinery repair, beauty shops, crafts, child care, and the like. In so far as the non-farm population has faced deteriorating economic conditions and declining opportunities in the formal sector in recent years, this study may offer a more general blueprint about how households cope with economic change.

The Restructuring of Farming in the 1980s

The 1980s crisis was the result of long-term structural trends and specific economic events. Historically, U.S. agriculture suffered from problems of overproduction, unstable and generally low farm incomes, and high government support costs (de Janvry and LeVeen 1986). Exportation of agricultural commodities was encouraged during the late 1960s as a strategy for dealing with these problems. In the process, however, farming became more vulnerable to new sources of price and income instability (de Janvry and LeVeen 1986). During the 1970s, exports expanded almost six times because of favorable economic conditions and the declining value of the dollar (Kolko 1988: 171). Farm prices rose and farmers expanded operations by investing in land, machinery and equipment; often they borrowed heavily to do so (Leistritz and Murdock 1988).

The conditions fostering agricultural expansion in the 1970s were reversed in the early 1980s. Farmers were faced with a national recession, a decrease in world demand for U.S. products due in part to the rising value of the dollar and the Russian grain embargo, and low commodity prices. Interest rates and the costs of producing farm commodities continued to rise. As returns to farmland diminished, land values declined: they fell 27 percent nationally from their 1981 peak, and almost 60 percent in some of the major farming states (Leistritz and Murdock 1988: 17). Declines in farm asset values were highest in those counties that specialized in export sensitive crops, wheat, soybeans, corn, cotton, and rice, the first three of which predominate in the Midwest (Gilles and Galetta 1993). Declining land and other farm asset values eroded farmers' equity. As a result, for some farmers, asset values fell below total liabilities and declining farm income made it difficult to service debt.

Farm liquidations increased as lenders reacted by refusing to extend credit to borrowers who appeared to be insolvent or by initiating legal action against farmers delinquent on loan payments (Leistritz and Murdock 1988: 17). Commercial farms defaulting on loans were concentrated in the North Central states with nearly 32,000 of such farms facing financial failure in the nation's hardest hit states of Iowa, Minnesota, and Wisconsin (Hanson 1990: 35).

Realization of the crisis was slow at first. Many people did not believe that the booming farm economy of the 1970s could enter a tailspin so suddenly. The first signs of crisis had been evident in the early 1980s, when real interest rates rose, farm exports plummeted, and land values declined. According to Harl (1990: 104), the gravity of these changes for farmers' financial well-being was not recognized widely until about 1984, in part because of the lack of up-to-date information about farm financial conditions and the distribution of debt. Land grant university and U.S. Department of Agriculture (USDA) researchers would reflect later on their shortcomings in collecting the data that could have provided earlier warning and more effective monitoring of the crisis. The media took notice of the crisis at about the same time as the academic community (Harl 1990: 222). Hardly an evening passed during the mid-1980s when stories of farm foreclosures, protests by farmers, and disrupted family life did not make headline news.

The federal government's recognition of the seriousness of the crisis came later. The first comprehensive response came through the Agricultural Credit Act of 1987. The Act provided financial assistance to the Farm Credit System and enabled farmers' loans to be restructured in lieu of foreclosure or informal liquidation. The dollar commitment by the federal government to the farm sector and its lenders exceeded $150 billion through the 1980s (Stam et al. 1991: 3). Without this massive infusion of aid from federal and state governments, substantially more persons would have exited farming.

Although much is known about the causes of the crisis, the magnitude of its effects is still unfolding, in part, because of the quality of existing data. A lack of detailed bankruptcy and foreclosure data, for instance, makes it impossible to determine precisely the number of forced exits from farming. By most accounts the crisis rivaled that experienced by farmers during the Great Depression. From 8 to 12 percent (200,000-300,000) of operators farming in 1980 are estimated to have failed financially by the end of the decade, becoming bankrupt, foreclosed, or financially restructured (Stam et al. 1991: 7).² Estimated forced sales of farmland jumped from 19 percent of all farmland transfers in 1980 to 46 percent in 1986 (Stam et al. 1991: 11). In the most severely affected areas in the Midwest, approximately one farmer in five faced the prospect of losing farm land and savings.

The crisis sent secondary shock waves to agricultural lenders and the input industry. More agricultural banks failed in 1987 than in any year since the Depression (Hanson 1990, 33). Nineteen billion dollars of bad loans (10 percent of all outstanding farm loans) were written off by agricultural lenders from 1984 to 1988 (Hanson 1990: 33). Farmers' capital purchases fell 60 percent from 1979 to 1986, contributing to the failure of International Harvester Corporation and the restructuring of other large agribusiness firms. Dunn and Bradstreet estimated that about 2,200 agricultural service firms failed (Hanson 1990: 33). Although it is recognized that the crisis spread beyond the sectors connected directly to farming and had tertiary effects on rural communities, most of the evidence for the latter is case-specific and anecdotal. Projected estimates by Murdock et al. (1988) suggest that by the mid-1990s, counties dependent on agriculture will experience large declines of the working age population, losses in wholesale and retail trade, local earnings, and tax revenues, reduced employment, and cuts in public services.

Because of the substantial popular, academic, and government attention given to the farm crisis, it tends to overshadow longer-term structural trends of the 1980s and the degree to which all farmers were affected. The magnitude of the crisis, however, should be understood in the context of historical patterns of farm change and its unevenness across the farm population. First, heavy restructuring of the farm sector had occurred already in the postwar decades preceding the crisis. Between 1980 and 1990, the number of farms declined by 12 percent and the farm population by 24 percent, the lowest rate of decline since 1950 (Stam et al. 1991: 21). Commercial farmers remaining in production during the crisis were generally efficient, profit-oriented, and technologically sophisticated—characteristics that enabled survival under highly competitive conditions. Prior epochs of farm restructuring had culled those who lacked these qualities.

Second, in a related vein, as is often noted, the crisis was particularly disheartening because it affected farmers most likely to succeed: well-educated younger persons who were considered innovative leaders (Stam et al. 1991: 27; Bultena, Lasley, and Geller 1986).

Third, the farm crisis seemed to accelerate the long term decline of moderate-sized family farms. Such farms are structurally less able to handle high debt loads than the largest farms, and they lack sufficient off-farm income to see them through hard times. Murdock et al. (1988: 157-158) estimate that about one-half of those who leave farming in counties dependent on agriculture from 1985 to 1995 will be family farmers from medium-sized farms.

Fourth, the characteristics of farm enterprises and people are highly varied, so that even farmers with similar types of enterprises and demographic backgrounds might experience it differentially. Some farmers obviously stood to gain by the lower cost real estate, machinery, and equipment from neighbors' exiting farming or downsizing operations.

Finally, the restructuring of the farm sector in the 1980s should be viewed as both specific to the crisis and as the continuation of postwar trends toward larger and fewer farms. Although involuntary exits from farming increased in the 1980s, the continued decline in the number of farms appears to be due more to the decreased entry of young farmers than to the increased exit of those remaining (Stam et al. 1990).

Perspectives on the Meaning and Significance of the Crisis

At least four distinct views of the farm crisis have emerged. The first attributes financial stress to farmers' personal characteristics and failings. That is, farmers who experienced the brunt of the crisis were viewed as poor and/or greedy managers who overextended themselves in pursuit of high profits and greater leisure time. This perspective was more common at the beginning of the crisis, and often is expressed in local accounts of failure. Farmers who deviated from traditional ideals about thrift and burdensome work, and who purchased modern machinery, upgraded their homes, and took vacations like urban people, were regarded as being punished for these attitudes or excesses.

Over time, as the crisis filtered through more segments of the farm population, charges of greed and poor management were debunked (Harl 1990). Occasionally, however, this "blaming the victim" approach is resurrected, particularly among those who survived the crisis. It provides a rationale for their own success and alleviates feelings of guilt and personal responsibility, founded or unfounded, about their neighbors' demise. Although farmers have a long-standing tradition of neighborliness and community solidarity, farming always has been a highly competitive business in which one person's loss is another's gain. Much of the farmland from those who exited during the crisis, for instance, was bought by other local farmers. Studies of those who failed showed they often were shunned by other farmers (Heffernan and Heffernan 1986; Rosenblatt 1990). Charges of greediness and poor management ability shift responsibility to the individual rather than to the government or the broader economic system.

According to a second perspective, the crisis was a unique occurrence in the development of an agricultural system that is basically socially responsible and economically viable (Harl 1990; Stam et al. 1991). The long-term trend toward larger and fewer farms, although dislocating some individuals, generally has resulted in net benefits to society, particularly in low-cost domestic food and more favorable balances of international trade. The crisis was due to an unusual combination of factors that the individual farmer could not fully control. These include the specific economic events of the 1980s, the period of time in which the operator entered farming, and the type of farm unit operated. Those who entered in years closer to the crisis, for example, were faced with higher costs for farm real estate and machinery. Being younger and at earlier stages in the life cycle, they tended to have less money in savings and other collateral. Cash grain farming, much of which takes place on moderate-sized Midwestern farms, was hit particularly hard due to declines in exports in the 1980s.

From this perspective, the crisis could be resolved through changing farm management strategies and through the regulatory structures that traditionally had guided agriculture, albeit with a massive infusion of capital. Farmers were cautioned to build up sizable equity, to decrease costs, and to diversify with livestock and off-farm employment (Hanson 1990). The federal government provided income support through commodity programs, new credit assistance programs, and enhancement of legal rights for borrowers. The percentage of farm revenue from total government support to the farm sector (including direct payments, credit, input and marketing subsidies, and benefits of legislation and research) doubled from 17.3 percent in 1982 to 35.8 percent in 1986 (Stam et al. 1991: 46). Although the bailout was costly, government programs and policies were temporarily successful. By the late 1980s, farm incomes and land values had begun to rise, a signal to some people that the crisis had ended and that the U.S. agriculture system was back on track (Stam et al. 1991: 3).

It is generally agreed that the crisis occurred for reasons beyond farmers' immediate control and that government aid to the farm sector was extremely helpful. This perspective, however, is limited in that it focuses almost entirely on the farm. Many analysts also would maintain that it underestimates the degree to which the farm crisis signaled more serious problems in the agricultural system and rural communities. The following perspectives argue that the crisis illustrates the need for fundamental change in government policy, although they advocate contrasting strategies to achieve this.

The third perspective views the farm crisis as a more serious problem of the effects of over-regulation of government. Government programs and policies have tended to benefit the largest farms creating inequities and making entry difficult for new, more efficient competitors. Government programs also have tended to reduce risk, thereby supporting operations that otherwise would not be viable. In effect, federal outlays serve as a form of welfare for the least competitive producers. At the same time, the rising federal deficit has made it ever more difficult to support traditional programs. Although in the short run the elimination or severe restriction of government aid would have real human costs, in the long-term farmers themselves and consumers would gain (Bovard 1989; Luttrell 1989).

The fourth perspective considers the farm crisis in the context of broader restructuring of agriculture and the rural economy of the 1980s (Davidson 1990; Kenney et al. 1989; Lobao 1990). Issues raised by this perspective include the growth of inequality and concentration in the agricultural system, the sustainability of present production practices, and the relationship between farming and the domestic and global economy. These issues were largely visible in prior postwar decades, but the financial solvency of the farm sector tended to obscure their gravity. The crisis became the tip of the iceberg: analysts no longer could ignore the serious problems facing farming.

Farming in the postwar era has been characterized by a rapid expansion of production made possible by capital-intensive technologies. It is premised on the production of low (direct)-cost food commodities using extensive natural resources, agrochemicals, and machinery. Some analysts refer to this system of farm production as "Fordist," a term applied to describe capital-intensive production of mass consumer goods epitomized first by the auto industry and later by other postwar industry (Kenney et al. 1989). In this system, farmers themselves have been sandwiched between "a monopoly controlled input sector and a monopoly controlled output sector," as evidenced by the concentration of large firms in industries such as agricultural chemicals, farm machinery and equipment, and food processing and distribution (Havens 1986: 45). The expansion of production and the increased need for borrowed capital have made farmers increasingly vulnerable to conditions outside the farm sector such as interest rates, the prices for inputs, and world market conditions.

This post-war economic environment also has witnessed the concentration of sales in a few large farms. Moderate-sized family farms have been less able to compete, and operators who wish to remain in farming must either take off-farm employment or expand production. Thus, greater inequality exists among segments of the farm population in terms of size of unit, ability to engage fully in farming, access to productive resources, and farm and family income.

During the 1980s, analysts also became more sensitized to other, related issues. The growing environmental movement increased public debate over food safety, soil erosion, groundwater contamination, and pesticide use. The use of highly capital-intensive production techniques requiring costly off-farm inputs called into question the long-term economic sustainability of the current farming system. Concern with the farm crisis further turned attention to the rural community as analysts recognized that viability of farming was interrelated with its local context. In doing so, they became aware of a broader rural "crisis," stemming from downturns in other traditionally rural industries and exacerbated by New Federalist programs and policies such as deregulation of transportation and banking.

From this perspective, traditional agricultural programs are insufficient for dealing with rural development. Rather, a comprehensive and sustained rural development initiative is necessary which would address issues such as employment, education, and health. Farm programs would be decoupled from initiatives proposing to strengthen rural well-being (Browne et al. 1992; Swanson and Skees 1987) and would be targeted to the moderate-sized family farms who needed them. According to proponents, such a comprehensive approach is ripe for political action for a number of reasons (de Janvry et al. 1986). Costly farm programs have become unbearable because of the national deficit. In addition, farm policy no longer can be used ideologically to justify protection of family farming when the evidence points to the unbridled growth of large-scale farming. In sum, this perspective argues that the postwar farm sector has witnessed fundamental changes which no longer could be ignored in the face of the crisis. Government programs and policy must undergo correspondingly deep alterations.

The economists and sociologists collaborating on this volume did not come to study farm restructuring from a uniform perspective. Economists generally have a longer tradition of working within the institutional framework of the U.S. agricultural system. They are more likely to view the present system as having evolved over time to fit the needs of consumers and of most farm people. Although they recognize the real human costs of farm restructuring, they point to changes that individual farmers and government can make to alleviate these costs. In contrast, sociologists tend to be less accepting of the parameters of the current agricultural system. They view the process of farm restructuring as generally destabilizing to social life and as contributing to greater inequality within the farm sector. With regard to policy, they are more likely to advocate fundamental changes in how society allocates its resources across different production units and segments of the population. Meaningful initiatives to counter the effects of farm restructuring must emerge not only from government but also from political action by stakeholders including farmers, farm workers, the environmental movement, and concerned public. Although we have noted contrast in the positions of economists and of sociologists, individuals from the two disciplines may overlap in their perspectives and may merge them into an eclectic view.[3]

Sociologists and economists thus generally came to study the farm crisis from different intellectual traditions. Their disciplinary training influences their assumptions about the magnitude and directions of farm change and about the government and popular initiatives needed to address it. The collaborative research on which this book is based was marked by economists' and sociologists' mutual respect for their different positions. Each discipline was viewed as illuminating in distinct ways the issues raised by the crisis and their potential policy solutions.

Focus of the Book: The Dimensions of Change in Farm Women's and Men's Lives During the Crisis

The purpose of this volume is to examine the consequences of the crisis for Midwestern farm women and men focusing on the areas of their lives which changes were likely to occur. In doing so, we also seek to address questions that remain generally unanswered or contested by analysts of the crisis.

Despite the attention to the causes of the farm crisis, its

consequences have not been comprehensively explored for broad segments of the farm population. First, as noted earlier, national data sources that could provide insight on the consequences of the crisis are limited. The decennial population census and the agricultural census makes it possible to examine changes only at a high level of aggregation such as the county. Some of the more detailed 1990 Census files still are not available at the time of this writing. Census public use samples of individuals typically do not collect data in sufficient detail to link enterprise characteristics with outcomes for individual or households.

Second, a lack of detailed bankruptcy, foreclosure, and other data in the 1980s makes it difficult to assess the severity of farmers' financial problems. The Bankruptcy Reform Act of 1978 terminated collection of data that would have separated farm from non-farm businesses (Stam et al. 1991: 9). Even if these data were available, it would be difficult to determine the extent to which financial stress pressured farmers to sell or transfer land and other assets to avoid bankruptcy or foreclosure. In addition, the farm population has become increasingly elderly; it is unclear to what extent financial pressure shaped plans for an earlier retirement.

Third, nearly all studies about farm financial stress during the 1980s were independent data collection projects conducted at state or lower administrative units. The results are not clearly generalizable across broad regions of the country, nor can direct comparisons be drawn between most of the studies.

Finally, as noted earlier, assessment of the effects of the crisis is complicated by the fact that studies focus mostly on financial impacts for the farm unit and only infrequently on the social and personal lives of the operator, spouse, and other household members. Information about how the crisis affected individuals' quality of life, gender roles, mental health, and socio-political attitudes remains largely anecdotal.

The restructuring of farming in the 1980s can be viewed as a process that transformed three dimensions of farm men's and women's lives, the focus of our study. Farming as a production system was disrupted by financial crisis as evidenced in the rise in debt levels, lowered land values, and in declining prices for agricultural commodities. Also, the Midwest experienced a severe drought during our period of study. In addition to these period shifts, farming continued to follow long-term, postwar restructuring trends. The farm household was also subject to change. Household members potentially faced deteriorating financial positions and declining consumption levels, increased social-psychological stress and depression, and adjustments in life-style, including the gender division of labor. Finally, changes were noted in the social and economic fabric of rural communities affected most severely by the crisis. Case studies portrayed the disruption of social networks, neighboring, and feelings of community solidarity (Davidson 1990; Heffernan and Heffernan 1986; Rosenblatt 1990). As farming deteriorated, so did other local businesses and community organizations and services.

Our focus on the farm enterprise, household, and community dimensions of farm women's and men's lives enables us to assess the major arenas of change during the farm crisis. In addition, previous popular as well as academic literature has raised questions about changes in these same dimensions to which the studies in this volume respond.

The Farm Enterprise: Financial Status,
Changing Production Strategies, and Future Needs

An important question is the financial status, including loan delinquency and rejection rates as well as solvency, of Midwestern farm enterprises at the close of the crisis. Farmers' financial status denotes both the effects of the crisis and the ability of farmers to withstand future shocks. While the crisis generally is regarded as having abated by the late 1980s, others have argued that crisis conditions continued and would set the stage for further declines in the 1990s (Davidson 1990). A related issue is that accounts of the crisis tend to discuss general trends giving the impression that financial distress in the Midwest was widespread. Although we know that Midwestern farmers tended to be affected more severely than those in other regions of the country due to their commodities and farm size, we question the extent to which most experienced severe financial stress.

A second set of questions deals with farmers' management strategies in response to change. Analysts suggest that those who survived the farm crisis did so by making a number of enterprise adaptations. Some of these responses are the types that farmers have made traditionally when facing hardship such as cutting costs, taking off-farm employment and quitting farming altogether (Bonanno 1987; Friedmann 1978). However, other adaptations portend a movement toward a new, "post-Fordist" agricultural structure: greater diversification of commodities and use of low-input sustainable practices, utilization of futures markets, and applying value-added techniques or on-farm processing of commodities (Kenney et al. 1989). To gauge the effects of the crisis, it thus is important to document adaptations of the 1980s as well as future plans which have implications for the newly emerging agricultural structure. The types of adaptations that farmers might make raise contrasting questions. Did the crisis function as a learning experience making farmers more cost conscious, market-aware, and sustainable stewards of farm land? Or alternatively, did a "business as usual" attitude return at the close of the crisis?

The Farm Household: Quality of Life, Division of Labor, and Social Psychological Outcomes

The farm crisis was widely considered to have affected household members' daily lives in a variety of ways. Perhaps the most common observation was that farm people's well-being declined. The agrarian ideal was belied by popular and academic accounts of household hardship, family adjustments, disenchantment with farming, and social-psychological distress. The pervasiveness of these portrayals, again, make hardship appear widespread. Thus, it is important to question the extent to which changes in quality of life varied among Midwestern farm households.

The expanding research and popular interest in farm women also raised questions about gender differences in the effects of the crisis. In other domestic and international contexts, traditional gender work roles have been set aside or renegotiated during periods of crisis. To what extent this occurred during the farm crisis is still unclear. The relatively few studies that have focused on work roles during this period center on off-farm employment. Findings from a census public-use record study show an increase of off-farm employment with distinct gender differences (Ollenburger, Grana, and Moore 1989). Farm women, relative to farm men, increased their participation in the labor force in the 1980s. Popular accounts, particularly in films such as *Country* and *The River*, highlighted women's increased participation in on-farm work as well. However, there is little scholarly information about the extent to which the traditional division of labor on the farm and household was altered. Did women increase farm work and broaden their base of production activities and decision making? Did men increase their share of domestic duties, particularly in light of women's apparent increase in off-farm employment?

A number of studies of the crisis period examined the link between economic hardship and farm people's social-psychological well-being such as stress and depression. Most were case studies (e.g., Heffernan and Heffernan 1986; Rosenblatt 1990) and tended to support such links. Quantitative studies over states, however, tended to find weaker evidence. For example, perceived rather than objective (debt-to-asset ratios, family income) financial distress was linked to social-psychological depression (Belyea and Lobao 1990) and coping and social support mechanisms were identified that modified the effects of financial distress (Armstrong and Schulman 1990; Lorenz et al. 1993). Few studies examined social-psychological outcomes by gender, although the distinct structural positions of farm women and men would be expected to lead to different coping strategies and mental distress. An exception is the just published volume of studies by Conger and Elder (1994) on rural Iowans, which includes farmers. Thus, a number of issues remain about the household and the gendered nature of crisis outcomes.

Community and Collectivity

By and large, studies of the farm crisis have focused on its most immediate, direct effects. Our most fragmented knowledge exists in what might be termed "third-order" effects, beyond those involving enterprise and household, and extending to community. These third order effects have been documented particularly in cases of collectivities undergoing hardship and change such as workers facing a local plant closure and communities in the midst of an environmental disaster (see Meyer, this volume). Studies on collectivities in crises have noted a variety of responses including shifts in previously held world views and community sentiments, increased personal stress, and, depending on local leadership and context, political action to counteract change. Popular and academic literature on the farm crisis suggests that all of these responses may have occurred.

As noted earlier, a variety of secondary government data show an increasing gap in social and economic conditions between metropolitan and non-metropolitan counties in the 1980s (O'Hare 1988). The farm population thus potentially witnessed decline in their communities during the crisis period. Poorer conditions in farming and in non-farm employment also tend to have multiplier effects that extend to public services, retail trade, and other aspects of community social organization. This was generally supported by case studies (Heffernan and Heffernan 1986; Rosenblatt 1990) and by the use of county-level estimates of changes that would occur in farming-dependent areas (Murdock et al. 1988). However, the extent to which farm people themselves observed such declines is unclear. There are no broadly generalizable studies that tap farm people's perceptions of changes in their communities during the crisis.

Farm people's political and organizational response to the changes of the 1980s has not been fully explored. Nearly all the existing scholarly information on this topic is from case study accounts. Two contrasting views emerge. The first is that there was heightened political activism among the farm population. Throughout the crisis years, dramatic events such as protests over farm foreclosures, the agitation of farm organizations, and standoffs between farmers and local law officials made media headlines. Farm women often were observed to be at the fore of this mobilization, which was underscored by the rise of women's independent farm organizations and by media accounts of individual women's courageous political acts. A second, contrasting perspective, however, is that farm women and men who became politically active were the exception. Further, farmers who experienced the brunt of the crisis likely would be less prone to activism due to factors such as their lack of financial resources and increased on- and off-farm work.

Exploring the Unevenness of Crisis and Change

The issues above raise a number of fundamental questions about the transformations in three arenas of farm people's lives. Debates about the crisis period remain, in part, because analysts have tended to treat Midwestern farmers as a homogenous group, to view hardship as widespread, and to draw conclusions from case studies, popular accounts, and other limited data sources. The studies in this volume question the extent to which transformations of the 1980s affected Midwestern farmers uniformly. They explore the unevenness of the crisis for farmers in different sub-regions, social class positions, and levels of enterprise commoditization. They also examine whether the crisis conferred differential experiences, perceptions, and behavioral outcomes for operators and spouses.

The use of a single survey of farm men and women across the Midwest enables us to address many of the questions that have been raised about the crisis. However, we realize that others remain. As will be shown in the following section, each dimension—the farm enterprise, the household, and the rural community—has been studied in a large literature and by certain academic specialties. More distinct focuses and questions pertaining to these bodies of literature obviously remain.[4]

The Dimensions of Change During the Crisis Period: A Conceptual Framework

The current research in economics and sociology allows for the development of a conceptual framework that can guide the study of farm restructuring and the questions posed above. This framework outlines the routes by which farm change is interrelated with changes in the household and community, thereby illustrating how the consequences of the farm crisis could diffuse through different dimensions of farmers' lives and segments of the farm population. We draw from the distinct literatures on farm, household, and community to which both economists and sociologists have contributed.

During the 1980s, the delicate balance between farm, household, and community life was disrupted. The farm enterprise is connected inextricably to the farm household so that a change in either becomes visible in both. The enterprise and the household also are embedded deeply in the rural community: changes in the rural community affect the character and survival of the farm enterprise and household. Likewise, communities that depend upon their farming hinterland are affected whenever change occurs in this sector. These interrelationships between farm, household, and community are essential to understanding the restructuring process (see Figure 1.1).

Farm Production and the Household

The farm household generally has been considered as a unit of both production and reproduction. The character of work and family decision making spills over into production and consumption; thus, it is both difficult and inappropriate to analyze changes in the farm enterprise apart from those in the household. Social scientists, however, have emphasized different "causal" directions of this relationship.

The effects of changes in farming such as declining and/or unstable income and high levels of debt on the household have been examined by sociologists, home economists, and other social scientists whose substantive focus is the rural family (Barlett 1993; Coward and Smith 1981; Lorenz et al. 1993; Marotz-Baden, Hennon and Brubaker 1988;

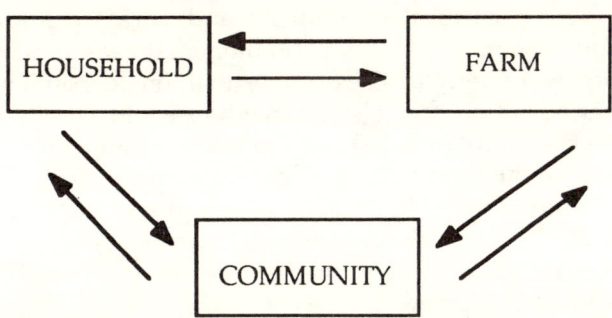

FIGURE 1.1 Dimensions of change during the crisis period

Rosenblatt 1990; Salamon 1992). These analysts view changes in the farm enterprise as altering household and personal characteristics such as quality of life, household consumption patterns and general survival strategies, and the balance of family power arrangements. Decline in income from farming is expected to result not only in declining family economic conditions but to increased levels of social-psychological stress and depression and marital disruption.

Farm families have been observed to employ a number of adaptive strategies to deal with financial pressures. Typically, they reduce household consumption, cutting back on food, transportation, and other expenditures (Bultena, Lasley, and Geller 1986). Studies of multiple job holding in farming demonstrate how inadequate farm income necessitates the need for off-farm employment if expansion of production is not feasible (Bonanno 1987). Households may piece together formal sector employment with informal work such as baby-sitting, roadside marketing of commodities, and crafts.

Changes in the farm enterprise have implications for the status and roles of women, now documented in large literature (Fink 1992; Haney and Knowles 1988; Rosenfeld 1985; Sachs 1983; Whatmore 1990). Farm women historically have had little ownership and control of factors of production such as land, capital, and technology. As farms increase in size, women tend to experience greater marginalization in regard to decision making and certain farm production activities such as field work. As farm income deteriorates (or as farm size decreases), farm women often balance more roles (farm, household, off-farm work) and have less leisure time than their partners. Probably because women are responsible for household reproductive activities, they tend to report a greater number of household adjustments and more perceived economic hardship (Lobao and Meyer 1991).

The obverse relationship, the effects of household characteristics on the dynamics of the farm enterprise, also has been examined primarily by agricultural economists and rural sociologists. The mobilization of family labor is a fundamental reason why farming remains largely a family business rather than one dependent on wage labor. In periods of economic stress, the household may "self-exploit": members may take off-farm work, may reduce consumption, or may work longer hours on the farm with less help (Friedmann 1978). The flexible nature of family production makes it possible to survive economic downturns albeit often with economic and emotional costs. In this sense, household strategies are a principal reason why family farming persists in advanced societies.

Other household related factors such as family size and life cycle, gender composition, and educational attainments affect how labor is allocated to farm work and how farm decision making occurs (Rosenfeld 1985). These factors thus affect structural features of the farm. For example, the decision to engage in part-time rather than full-time farming is, in part, a function of members' human capital and life-cycle. In general, men with higher education and prior off-farm job experience are more likely to farm part-time. Similar human capital factors operate for farm women although the presence of young children may be more likely to impede their labor force participation (Godwin and Marlowe 1990). As noted previously, the relationship between farm and household are reciprocal, so that a decision to engage in part-time farming, in turn, stands to alter other household arrangements.

The Farm, the Household, and the Rural Community

The farm enterprise and household also are tied closely to the character of the rural community. Changes in farming reverberate beyond the immediate household affecting other local people and businesses and the quality of community life. A long literature from agricultural economics and rural sociology has addressed the effects of farm change on aggregate community well-being (Lobao 1990). Much of this has been aimed at testing Goldschmidt's (1978) hypothesis that community well-being is jeopardized by the growth of fewer and larger farms, or concomitantly, by the decline of moderate-size family farms. Such changes in farm structure have been reported to lead to declines in local population, lower standards of living, less community organizational participation and social integration, decreased retail trade, and greater unemployment. However, according to recent studies, the relationship between farm change and community well-being is not so clear cut. The effects of farm change have been found to vary over time and by region of the country, presence of local non-farm employment, human capital, and other factors (Lobao 1990; Lobao and Schulman 1991; Swanson 1988; Green 1985).

Surveys of the farm population have also examined how structural change, evident in farmers' class locations, affects their community participation and sense of attachment. In general, moderate-sized family farmers have been found to have greater participation in community activities and feelings of community integration in contrast to farmers lower in the class structure (Heffernan 1972; Heffernan and Lasley 1978; Martinson, Wilkening, and Rodefeld 1976; Poole 1981).

The few studies that have addressed the effects of farm change on community well-being during the 1980s crisis are concerned with the loss of middle class, independent producers who sustain community institutions and businesses. Heffernan and Heffernan (1986), for example, in studying social psychological depression among family farmers argued that communities may undergo a corresponding collective depression which inhibits new economic investment and impairs the vitality of social life. Murdock et al. (1988) estimate that the long-term effects of the crisis in agriculturally-dependent counties will be evident in strains in local social organization, community solidarity, and increased social-psychological problems.

The effects of the changing farm economy on political activism in the rural community and beyond is documented in a large literature. Much of this is grounded in case studies and historical accounts from the Whiskey Rebellion onward. A long-standing observation has been that farmers' political behavior tends to be more influenced by pocketbook considerations than that of other Americans (Campbell et al. 1960). More recent studies, however, have disputed this tendency (Sigelman 1987).

The community, in turn, circumscribes and affects both the farm enterprise and the farm household. It represents the context in which farm and household decision making occurs and household adaptations are made possible. The local context shapes the structure of farming in a number of ways, a topic addressed by both agricultural economists and rural sociologists (Beaulieu, Miller, and Mulkey 1988; Korsching and Stofferahn 1986; Salant and Munoz 1981). Ecological characteristics, including the agricultural commodity region, affect farm structure. For example, larger farms tend to be located to the west in the wheat-growing region of the North Central plains, and smaller farms to the east in the corn belt. The non-farm economy also affects farm structure through land, labor, and product markets. In more urbanized areas, the high cost of land affects the number and size of farms. The presence of paid employment opportunities facilitates part-time farming and may serve to raise wages of hired farm workers. Local businesses supplying farm inputs or engaged in marketing and processing of farm outputs can help to sustain smaller farms because these are less likely than large farms to be integrated into non-local supplier-processor networks.

Other aspects of community context have a more direct bearing on the farm household. Services such as retail stores, banks, schools, and hospitals, and infrastructure such as roads and utilities are obviously important to the consumption needs of the farm household. There is some evidence that because services and infrastructure facilitate the reproduction of farm households, family farming is more likely to persist in locations where these are better (Green 1985; Lobao 1990). The character of the local labor market affects household members' decisions such as allocation of time between leisure and work. Factors such as job availability and quality of jobs, distance to employment, population density, and the amount of job sex-segregation are commonly cited as influencing individuals' labor force participation and earnings (Findeis, Lass, and Hallberg 1991; Godwin and Marlowe 1990).

Organization of This Book

The studies in this volume build on the previous framework and recognition of the reflexive nature of changes between the farm, the household, and the community. Because our study centers on the survivors of the crisis, our understanding about changes in these three dimensions of life is drawn from farm household members, specifically the operator and spouse. We are concerned with the farm and household transformations that occurred in the wake of restructuring of the 1980s and with the attitudinal and emotional experiences these engendered. The community is considered in terms of how farm people perceived it and how it provides order to their lives rather than through the secondary data analysis of quantitative indicators. Thus, our study provides insight into the extent to which the local community was transformed from the vantage of those tied most closely to the community through the agrarian way of life. Our focus on the farm, household, and community, coupled with the use of a single survey of farm men and women across the Midwest, enables us to address many of the questions that have been raised about the crisis in regard to changes in the dimensions of farm people's lives.

The present chapter provides an overview of the substantive focus, research questions, and conceptual issues elaborated in chapters to follow. The North Central regional farm survey, on which findings from this volume are based, is described in the Appendix. Chapter 2 by Hobbs and Weagley, provides a broad overview of the changes in farming and rural life that occurred in the North Central region during the 1980s. The subsequent chapters are organized into three sections.

Part I centers on the direct outcomes of farm restructuring for the enterprise. The authors examine the financial status of the regions' operations (Chapter 3), production adaptations and consequences (Chapter 4), and future plans for the business organization and management training needs (Chapter 5). These chapters describe how the crisis unfolded in different sub-regional locations and for different types of enterprises. They show that the effects of the crisis were highly uneven but served as a specter for many farmers to make enterprise and occupational adjustments.

Part II examines the response to farm change by the household, operator, and spouse. The authors discuss the extent to which the division of farm, off-farm, and household labor were altered (Chapter 6), and the effects of financial hardship on household consumption and well-being (Chapter 7). Again, the consequences of farm restructuring for household well-being are shown to be highly uneven. Findings with regard to the division of labor indicate that spouses were more likely to take off-farm employment. However, popular observations about other changes in the traditional division of household and farm labor generally are shown to be unfounded. The extent to which farm spouses and operators experienced and coped with farm change differentially is explored in a chapter on social-psychological outcomes (Chapter 8).

Part III focuses on the less direct effects of restructuring, beyond enterprise and household. The extent to which farm people in different sub-regional locations perceived community change is explored (Chapter 9). Women's and men's participation in community organizations and their mobilization into protest and forms of political activism is documented (Chapter 10). These chapters overturn common generalizations about the crisis period and/or delineate the specific segments of the farm population to which they might apply.

The final chapter (11) summarizes the meaning and significance of the findings particularly with regard to the future of family farming in the Midwest.

As noted previously, the authors of this volume do not share a single perspective as to the significance and policy implications of farm restructuring. Our goal was to assess empirically the changes that occurred during the 1980s for a large segment of the farm population. Because of our disciplinary traditions and specific topics of focus, differences of style, method, and theoretical explicitness persist. Nevertheless the questions raised and the answers suggested by the economists and sociologists included here are a matter of common interest. They also address some of the most fundamental public concerns about the trajectory of change in U.S. farming.

Notes

1. Definitions of a small, medium, large or family farm vary. Sales criteria are used most often to define farm size; a moderate-sized farm now has about $100,000-499,999 in sales. The definition of a family farm is also debated, but generally this refers to a unit in which the operator and his or her family control decision making, supply most of the labor, and own or control farm real estate and other assets. Some also differentiate traditional family farming or simple commodity production from part-time farming or pluriactivity on the basis that the household derives income from non-farm sources. Reasons vary for the decline of moderate-sized family farms, particularly during the post-war

period. See Lobao (1990) for a discussion of perspectives on why family farming has declined.

2. It should be noted that many of those who failed financially did not leave the farm sector entirely. However, many operations that continued were at a greatly reduced scale (Stam et al. 1991: 7).

3. Recently, a number of attempts have been made to open the borders between sociology and economics and to employ insights from both disciplines to analyze human behavior (Swedberg and Granovetter 1992). Most of these hybrid efforts are still in an embryonic stage. Also debated is whether a holistic theory of human economic behavior from a synthesis of sociology and economics can be built or even is desirable. Rather, our position follows Swedberg's (1991) more modest suggestion that multiple approaches to economic problems should be taken and that interdisciplinary borders should be kept open. Allowance should be made for the complexity of human behavior and for the effects of social structure and culture.

4. The purpose of this study is to assess the consequences of the 1980s farm restructuring for Midwestern women and men, rather than to provide an exhaustive treatment of the farm enterprise, the family, and the rural community.

References

Armstrong, Paula S. and Michael D. Schulman. 1990. "Financial Strain and Depression among Farm Operators: The Role of Perceived Economic Hardship and Personal Control." *Rural Sociology* 55 (Winter): 475-493.

Barlett, F. 1993. *American Dreams, Rural Realities: Family Farms in Crisis.* Chapel Hill: The University of North Carolina Press.

Beaulieu, Lionel J., Michael K. Miller, and David Mulkey. 1988. "Community Forces and Their Influence on Farm Structure." Pp. 211-232 in Lionel J. Beaulieu (ed.), *The Rural South in Crisis: Challenges for the Future.* Boulder: Westview.

Belyea, Michael J. and Linda M. Lobao. 1990. "Psychosocial Consequences of Agricultural Transformation: The Farm Crisis and Depression." *Rural Sociology* 55 (1): 58-75.

Bonanno, Alessandro. 1987. *Small Farms: Persistence with Legitimation.* Boulder: Westview.

Bovard, James. 1989. *The Farm Fiasco.* San Francisco: ICS Press.

Browne, William P., Jerry R. Skees, Louis E. Swanson, Paul Thompson, and Laurian Unnivher. 1992. *Sacred Cows and Hot Potatoes: Agrarian Myths in Agricultural Policy.* Boulder: Westview.

Bultena, Gordon, Paul Lasley, and Jack Geller. 1986. "The Farm Crisis: Patterns and Impacts of Financial Distress among Iowa Farm Families." *Rural Sociology* 51 (Winter): 436-448.

Campbell, Angus, Philip Converse, Warren Miller, and Donald Stokes. 1960. *The American Voter.* New York: Wiley.

Carlin, Thomas A. 1990. "An Overview." Pp. 1-6 in *The U.S. Farming Sector Entering the 1990s: Twelfth Annual Report on the Status of Family Farms*. Agriculture Information Bulletin 587. Washington, DC: U.S. Department of Agriculture, Economic Research Service.

Conger, Rand D. and Glen H. Elder, Jr. 1994. *Families in Troubled Times: Adapting to Change in Rural America*. Hawthorne, NY: Aldine de Gruyter.

Coward, R. T. and W. Smith, Jr. (eds.). 1981. *The Family in Rural Society*. Boulder: Westview.

Davidson, Osha Gray. 1990. *Broken Heartland: The Rise of America's Rural Ghetto*. New York: Anchor Books.

de Janvry, Alain and Phillip LeVeen. 1986. "Historical Forces That Have Shaped World Agriculture: A Structural Perspective." Pp. 83-104 in *New Directions for Agriculture and Agricultural Research: Neglected Dimensions and Emerging Alternatives*, edited by Kenneth A. Dahlberg. Totowa, NJ: Rowman and Allanheld.

Findeis, Jill L., Daniel A. Lass, and Milton C. Hallberg. 1991. "Effects of Location on Off-Farm Employment Decisions." Pp. 263-282 in *Multiple Job-Holding among Farm Families*, edited by Milton C. Hallberg, Jill L. Findeis, and Daniel A. Lass. Ames: Iowa State University Press.

Fink, Deborah. 1992. *Agrarian Women: Wives and Mothers in Rural Nebraska, 1880-1940*. Chapel Hill: University of North Carolina Press.

Friedmann, Harriet. 1978. "World Market, State, and Family Farm: Social Bases of Household Production in an Era of Wage Labor." *Comparative Studies in Society and History* 20: 545-586.

Gilles, Jere L. and Simon Galetta. 1993. "Is Trade Always Good for Rural America." Unpublished manuscript. Columbia, MO: Department of Rural Sociology.

Godwin, Deborah and Julia Marlowe. 1990. "Farm Wives' Labor Force Participation and Earnings." *Rural Sociology* 55 (1): 25-43.

Goldschmidt, Walter. 1978. *As You Sow: Three Studies in the Social Consequences of Agribusiness*. Montclair, NJ: Allanheld, Osmun, and Company.

Goss, K., Rodefeld, R. and Buttel, F. 1980. "The Political Economy of Class Structure in U.S. Agriculture." Pp. 83-132 in *The Rural Sociology of The Advanced Societies*, edited by F. Buttel and H. Newby. Montclair, NJ: Allanheld, Osmun, and Company.

Green, Gary P. 1985. "Large-scale Farming and the Quality of Life in Rural Communities: Further Specification of the Goldschmidt Hypothesis." *Rural Sociology* 50 (Summer): 262-273.

Haney, Wava G. and Jane B. Knowles. 1988. "Making the 'Invisible Farmer' Visible." Pp. 1-12 in *Women and Farming: Changing Roles and Changing Structures*, edited by Wava G. Haney and Jane B. Knowles. Boulder: Westview.

Hanson, Greg. 1990. "Beyond the Farm Debt Crisis." *Choices* (Fourth Quarter): 33-35.

Harl, Neil E. 1990. *The Farm Debt Crisis of the 1980s*. Ames: Iowa State University Press.

Havens, A. Eugene. 1986 "Capitalist Development in the United States: State Accumulation, and Agricultural Production System." Pp. 26-59 in *Studies in the Transformation of U.S. Agriculture*, edited by A. Eugene Havens with Gregory Hooks, Patrick H. Mooney, and Max J. Pfeffer. Boulder: Westview.

Heffernan, W. D. 1972. "Sociological Dimensions of Agricultural Structures in the United States." *Sociologia Ruralis* 12 (October): 481-499.

Heffernan, W. D. and Paul Lasley. 1978. "Agricultural Structure and Interaction in the Local Community: A Case Study." *Rural Sociology* 43 (Fall): 348-361.

Heffernan, W. D. and Heffernan, J. B. 1986. "The Farm Crisis and the Rural Community." Pp. 273-280 in *New Dimensions in Rural Policy: Building Upon Our Heritage, Studies Prepared for the Use of the Subcommittee on Agriculture and Transportation of the Joint Economic Committee of the United States*, edited by D. Jahr, J. W. Johnson, and R. C. Wimberley. Washington, DC: U.S. Government Printing Office.

Kenney, Martin, Linda M. Lobao, James Curry, and W. Richard Goe. 1989. "Midwestern Agriculture in U.S. Fordism: From New Deal to Economic Restructuring." *Sociologia Ruralis* 29 (2): 130-148.

Kolko, Joyce. 1988. *Restructuring the World Economy*. New York: Pantheon Books.

Korsching, Peter F. and Curtis W. Stofferahn. 1986. "Agricultural and Rural Community Interdependencies." Pp. 245-266 in *Agricultural Change: Consequences for Southern Farms and Rural Communities*, edited by Joseph Molnar. Boulder: Westview.

Krause, K. 1987. *Corporate Farming, 1969-1982*. Agricultural Economic Report Number 578. Washington, DC: U.S. Department of Agriculture, Economic Research Service.

Leistritz, F. L. and Murdock, S. H. 1988. "The Implications of the Current Farm Crisis for Rural America." Pp. 13-28 in *The Farm Financial Crisis: Socioeconomic Dimensions and Implications for Producers and Rural Areas*, edited by S. H. Murdock and F. L. Leistritz. Boulder: Westview.

Lobao, Linda M. 1990. *Locality and Inequality: Farm and Industry Structure and Socioeconomic Conditions*. Albany: State University of New York Press.

Lobao, Linda M. and Katherine Meyer. 1991. "Farm Restructuring, Adaptations in Household Consumption, and Stress among Farm Men and Women." Pp. 191-209 in *Research in Rural Sociology and Development: A Research Annual* Volume 5, edited by H. K. Schwarzweller and D. Clay. Greenwich, CT: JAI Press.

Lobao, Linda M. and Michael D. Schulman. 1991. "Farming Patterns, Rural Restructuring, and Poverty: A Comparative Regional Analysis." *Rural Sociology* 56 (Winter): 565-602.

Lorenz, Frederick O., Rand D. Conger, Ruth B. Montague, and K. A. S. Wickrama. 1993. "Economic Conditions, Spouse Support, and Psychological Distress of Rural Husbands and Wives." *Rural Sociology* 58 (Summer): 247-268.

Luttrell. 1989. *The High Cost of Farm Welfare*. Washington, DC: Cato Institute.

Marotz-Baden, Ramona, Charles B. Hennon, and Timothy H. Brubaker, eds. 1988. *Families in Rural America: Stress, Adaptation and Revitalization*. St. Paul, MN: National Council on Family Relations.

Martinson, Oscar B., Eugene A. Wilkening, and Richard D. Rodefeld. 1976. "Feelings of Powerlessness and Social Isolation Among 'Large Scale' Farm Personnel." *Rural Sociology* 41 (Winter): 452-472.

Murdock, Steve H., Lloyd B. Potter, Rita R. Hamm, Kenneth Backman, Don E. Albrecht, and F. Larry Leistritz. 1988. "The Implications of the Current Farm Crisis for Rural America." Pp. 141-168 in *The Farm Financial Crisis: Socioeconomic Dimensions and Implications for Producers and Rural Areas*, edited by S. H. Murdock and F. L. Leistritz. Boulder: Westview.

O'Hare, William P. 1988. *The Rise of Poverty in Rural America*. Occasional Paper Number 15, July. Washington, DC: The Population Reference Bureau.

Ollenburger, Jane C., Sheryl J. Grana, and Helen A. Moore. 1989. "Labor Force Participation of Rural Farm, Rural Nonfarm, and Urban Women." *Rural Sociology* 54 (4): 533-550.

Poole, Dennis L. 1981. "Farm Scale, Family Life, and Community Participation." *Rural Sociology* 46 (Spring): 112-127.

Rosenblatt, C. 1990. *Farming Is in Our Blood: Farm Families in Economic Crisis*. Ames: Iowa State University Press.

Rosenfeld, R. A. 1985. *Farm Women: Work, Farm, and Family in the United States*. Chapel Hill: University of North Carolina Press.

Sachs, Carolyn. 1983. *The Invisible Farmers: Women in Agricultural Production*. Totowa, NJ: Rowman and Allanheld.

Salamon, Sonya. 1992. *Prairie Patrimony: Family, Farming, and Community in the Midwest*. Chapel Hill: University of North Carolina Press.

Salant, Priscilla and Robert Munoz. 1981. *Rural Industrialization and Its Impact on the Agricultural Community: A Review of the Literature*. Washington, DC: U.S. Department of Agriculture, Economics and Statistics Service Staff Report.

Sigelman, Lee. 1987. "Economic Pressures and the Farm Vote: The Case of 1984." *Rural Sociology* 52 (2): 151-163.

Stam, J. M., S. R. Koenig, S. E. Bentley, and H. F. Gale, Jr. 1991. *Farm Financial Stress, Farm Exits, and Public Sector Assistance to the Farm Sector in the 1980*. Agricultural Economic Report Number 645. Washington, DC: U.S. Department of Agriculture, Economic Research Service.

Swanson, Louis E., ed. 1988. *Agriculture and Community Change in the U.S.: The Congressional Research Reports*. Boulder: Westview.

Swanson, Louis E. and Jerry R. Skees. 1987. "Finding New Ideas for Old Objectives: The Current Case for Rural Development Programs." *Choices* 4 (Fourth Quarter): 8-32.

Swedberg, Richard M. 1991. "The Battle of Methods: Toward a Paradigm Shift?" Pp. 13-33 in *Socio-economics: Toward a New Synthesis*, edited by Amitai Etzioni and Paul R. Lawrence. Armonk, NY: M. E. Sharpe, Inc.

Swedberg, Richard and Mark Granovetter. 1992. "Introduction." Pp. 1-26 in *The Sociology of Economic Life*, edited by Mark Granovetter and Richard Swedberg. Boulder: Westview.

Whatmore, Sarah. 1990. *Farming Women: Gender, Work and Family Enterprise*. London: Macmillan Academic and Professional Ltd.

2

The Agricultures of the Midwest and Their Demographic and Economic Environments

Daryl Hobbs and Robert Weagley

Since the end of the 1970s, the nation has witnessed rural life in transition. This has been a movement of capital, factor costs, and people. Rural America has been changing from a robust network of family farms and of rural businesses serving those farms to a patchwork of larger agricultural producers, processors, and suppliers dependent on increasingly centralized centers of rural commerce. The rural scene is occupied increasingly with communities and people struggling to preserve their economic futures.

This book does not analyze the factors that accelerated this shift during the 1980s. Rather, it documents the reactions of the people affected most directly by this evolution—the farm families. The socioeconomic conditions of the North Central region must be understood, however, as the context in which these families acted. To understand this context we must start with a description of the economic forces that affected the farm families and rural communities of the region: credit demand, interest rates, farm product prices, exports, and land values.

Credit indeed is a product of expectations, and in the 1970s rural expectations were bright. From 1975 through 1980, real rates of interest actually were negative (see Figure 2.1); as a result, farmers were eager to borrow money. They used these funds not only to expand their operation and gain the advantages of increasing economies of scale, but

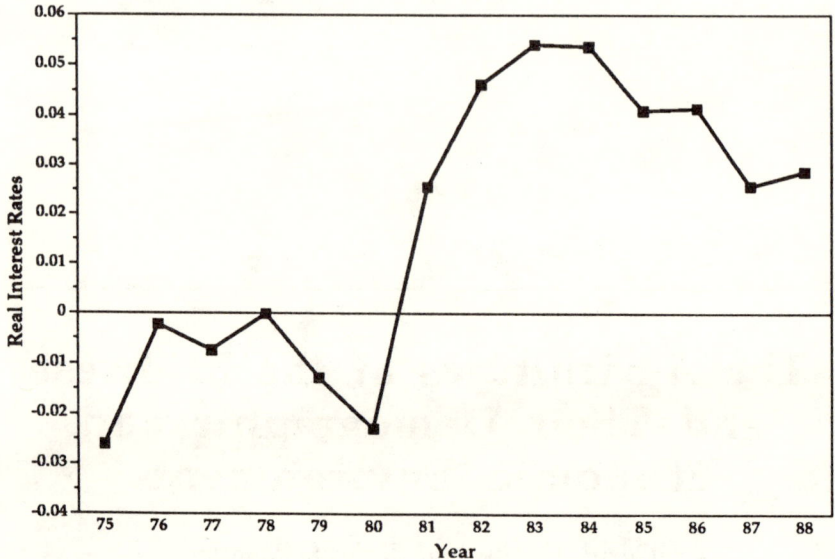

FIGURE 2.1 Real interest rates 1975 through 1988

to also increase production and thereby capture the relatively high prices for agricultural commodities that existed during the late 1970s. For the period 1975-1988, prices for wheat, corn, and pork in 1975 dollars peaked in 1975 at $3.55 and $2.54 per bushel and $46.10 per 100 pounds, respectively. In 1976 the prices of chicken, soybeans, and milk peaked at $0.122 per pound, $6.44 per bushel, and $9.13 per 100 pounds respectively; beef prices peaked in 1979 at $48.98 per 100 pounds (see Figures 2.2 and 2.3).

At the same time, it appeared that world food supplies depended on U.S. agricultural production. The value of the U.S. dollar declined in relation to both the German mark and the Japanese yen during the late 1970s; consequently the value of total agricultural exports peaked in 1980 (see Figure 2.4). These trends encouraged farmers to answer the call of then-Secretary of Agriculture Butz to "plant fence row to fence row" to feed the world and to capitalize on the boom in agriculture.

This desired expansion gained momentum as baby-boom children in the North Central region completed school, started their families, and began their farm operations, often with the benefit of equity in their parents' farm enterprise. All of these factors combined to send the price of farmland higher and higher, thus increasing further the equity of those who owned land and making it easier to gain access to agricultural credit markets. The prevailing wisdom paralleled the

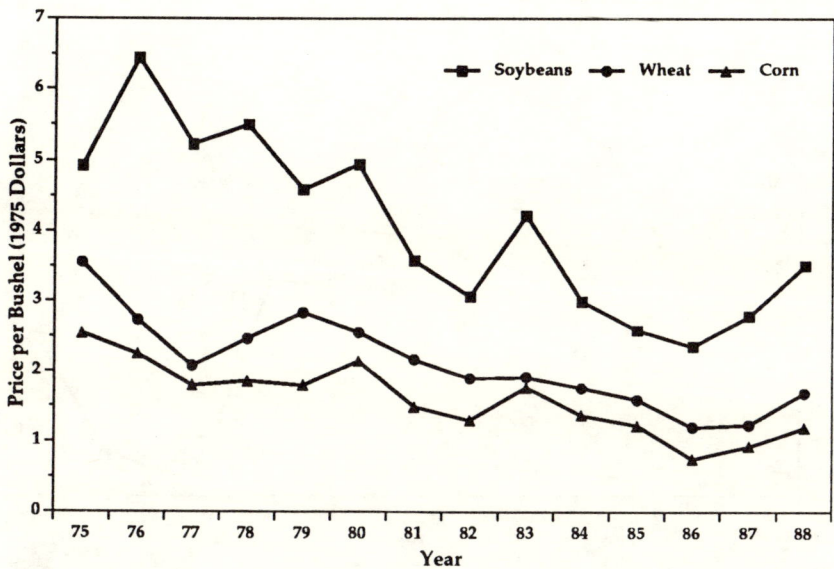

FIGURE 2.2 Soybean, wheat, and corn prices 1975 through 1988

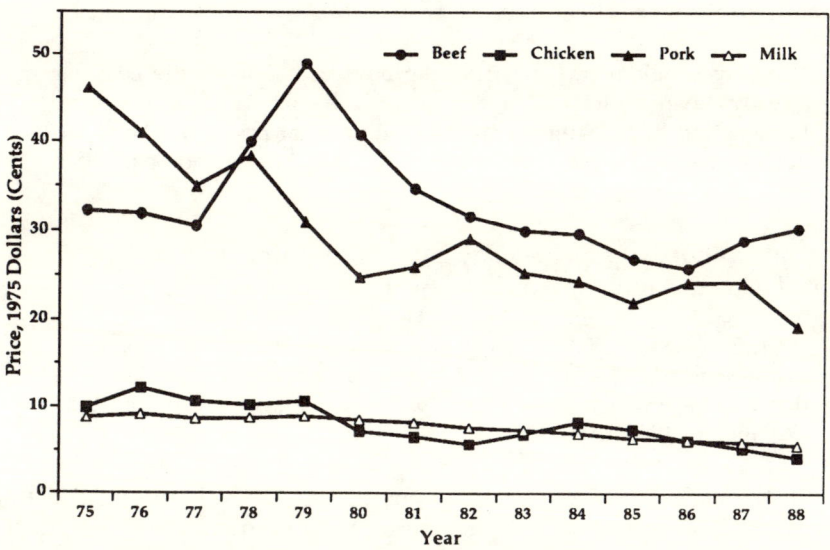

FIGURE 2.3 Beef, chicken, pork, and milk prices 1975 through 1988

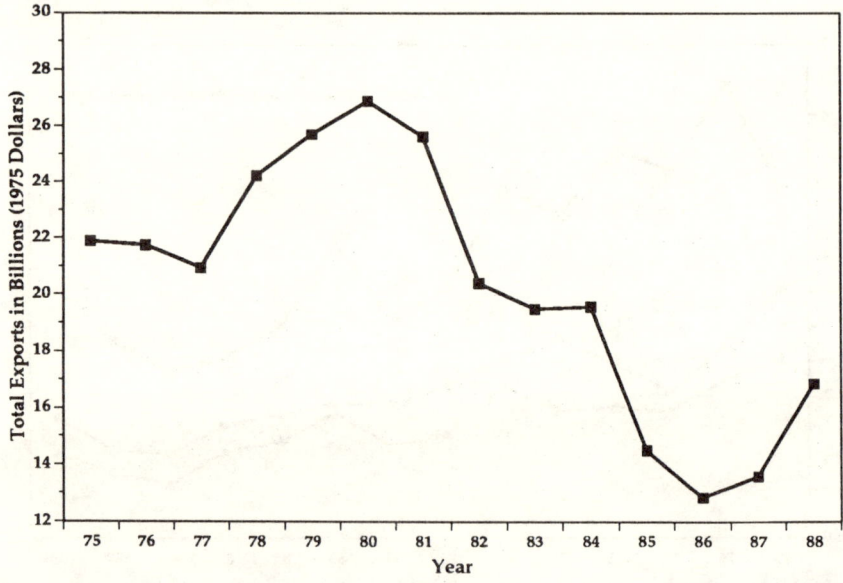

FIGURE 2.4 Total exports in billions 1975 through 1988 (1975 dollars)

thinking of the Iowa farmer quoted below, who spoke as follows in a *Farm Journal* cover article titled "Can You Make $3,000 Land Pay?"

> The way I look at it, I'm letting the boys use some of the equity I already have in the land. It'll be a while before I'm ready to quit farming and by that time, there's no telling what prices will be. If we don't buy now, we may not be able to at all (Schotsch and Seim 1980: 10).

Unfortunately, four months after publishing this article, *Farm Journal* quoted another Iowan, land auctioneer James J. McGuire:

> ...farms worth $3,000 an acre three to six months ago may have dropped to $2,000 since March 1—and may be worth even less. We don't know because we haven't been able to get acceptable bids (Wennblom 1980: 44).

In the early 1980s the expectations of the 1970s were eclipsed by the beginnings of a major restructuring of the family farm sector. Real interest rates went from negative to the highest level in years. These rates attracted foreign investors and the value of the dollar rose on international markets. Farm exports became increasingly noncompetitive in world markets, and the rising domestic supply caused farm product prices to fall sharply. Unfortunately, many overleveraged farmers fell with them.

Between 1980 and 1981 real interest rates became positive. They increased steadily to a period high of 5.4 percent in 1983 (Figure 2.1), while the market value of farm commodities declined (Figures 2.2 and 2-3). Increasing interest rates, coupled with declines in the prices of the products of the land, hastened the economic collapse of agriculture across the entire North Central region. As shown in Figures 2.5 and 2.6, average per-acre farmland values for 1975-1988 (in 1975 dollars) peaked in 1979 in Illinois ($1,324), Ohio ($1,123), and South Dakota ($190). In 1980 the average farmland values were greatest in Indiana ($1,197), Iowa ($1,182), Kansas ($374), Missouri ($573), Nebraska ($392), and North Dakota ($261); in 1981 the peak occurred in Minnesota ($729), Wisconsin ($654), and Michigan ($729).

In an attempt to save their farm business, many farm operators and spouses sought off-farm employment. During this period, however, rural manufacturing began to move overseas to employ less costly labor. Through 1985, these workers could be paid in increasingly lower-valued currencies, which worked in conjunction with the increasing supply of

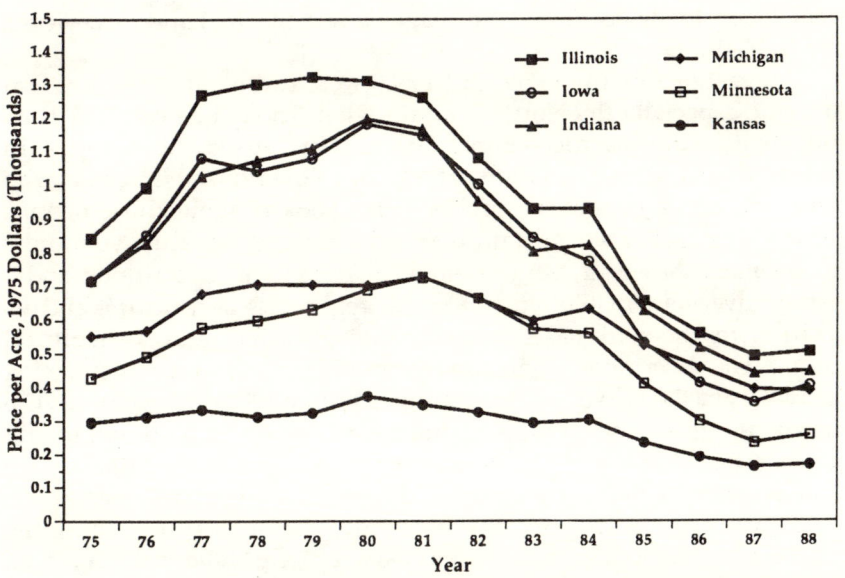

FIGURE 2.5 Average farm land values by state 1975 through 1988 (1975 dollars)

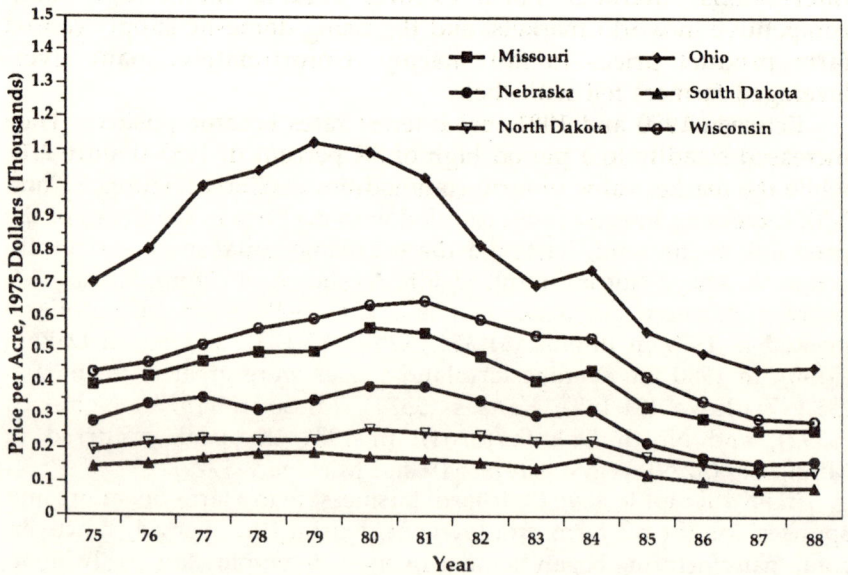

FIGURE 2.6 Average farm land values by state 1975 through 1988 (1975 dollars)

rural labor to drive down the wages available to rural Americans (Flora 1990).

Unquestionably the 1970s and 1980s were turbulent times for rural America, especially the North Central region. Today one can look back and identify the macroeconomic factors that caused the changes. In that backward glance, it is important not to overlook the people who were, and are, a part of the process. This book is about those people. Before discussing farm families' adaptations to this economic environment, however, we present a socioeconomic description and a historic overview of the North Central region, its agriculture, and its position in the national economic, demographic, and agricultural picture. The description is intended to provide a context for interpreting the results of the survey of farm families presented in this volume. The survey results show substantial variation in responses both within and between the 12 states of the region. A reason for this variation, as we shall describe, is that even though these states are grouped together as the North Central region, the economy, the agriculture, and the patterns of demographic change vary greatly. The variations in economic conditions and change are associated with different patterns of adaptation by farmers, as the following chapters report.

The Social, Economic, and Agricultural Restructuring of the Region

The 12 states of the North Central region (the Midwest) historically have been called the corn belt or, more recently, the farm belt. The region's role as breadbasket of the nation began a century ago and contributed directly to the commercial and industrial growth of some of its major cities, such as Minneapolis-St. Paul, Chicago, Kansas City, and Omaha. Carl Sandburg immortalized Chicago as meatpacker of the world; the Twin Cities gained a reputation as the nation's miller. Railroads connected the region's farms with the cities and facilitated the growth of an urban manufacturing complex. City-bound trains, loaded with grain or livestock, usually made the return trip to rural areas loaded with automobiles, machinery, building materials, and the like.

As the twentieth century proceeded, the region's farms became increasingly productive, largely by substituting capital in the form of machinery and technology for labor. Concurrently the region was industrializing. By mid-century the Midwest included not only the nation's corn belt but a substantial portion of its industrial belt as well. Heavy industry constituted the region's industrial base: the nation's production of automobiles, trucks, railroad cars, construction equipment, tires, and farm machinery became concentrated in the region. The output of both farms and cities expanded quickly; several of the region's states become as well known for their industry as for their agricultural production. As a result the region's educated farm youths, whose labor had been replaced by machinery on the farms, generally found a market for their skills and their work ethic in the industrializing cities. Correspondingly, throughout the first 60 years of the twentieth century, the population of the region's cities increased in direct proportion to population losses in rural areas. Indeed, many rural counties in the region achieved their peak population by the early 1900s and have suffered continuing losses since that time.

Over the past 25 years, however, substantial restructuring has occurred both on the farms and in the cities. Chicago, Kansas City, and Omaha are no longer the meat-packing centers of the nation, nor are those cities the rail hubs they once were. The manufacture of cars, trucks, and other durables has dispersed both nationally and globally, attracted by proximity to larger markets and by lower labor costs. This decline in the region's manufacturing base led to the term "rust belt," coined in the early 1980s to refer to the declining industrial base from St. Louis to Boston. The appearance of the "rust belt" coincided with the emergence of the "farm crisis." Together these dramatic changes produced serious erosion of the economic strength that the region had enjoyed during the earlier years of the century. During the past decade, however, economic restructuring has begun to revive the region's cities, while the countryside falls further behind.

Recent Population Changes

The economic restructuring of the North Central region has been reflected recently in the slowness of population growth in relation to the rest of the country. During the past three decades, the region's cities, like the farms and the rural communities, exported some of their population. The destination of those who departed was often the "sun belt" of the South and the West. During the 1960s, the region's population grew by 9.6 percent (13.4 percent for the United States); during the 1970s, population growth declined to 4.0 percent (11.4 percent for the United States). The rate of growth declined even further to 1.4 percent in the 1980s (compared with 9.8 percent for the United States).

The national shift in population from the Midwest and the Northeast to the South and the West is reported in Table 2.1. The population gains of the South and West occurred largely at the expense of the Midwest and Northeast as the result of substantial migration from those regions. From 1960 to 1990 the population of the Northeast and the Midwest grew by only 13.7 and 15.6 percent respectively, while the South increased by 55.4 percent and the West by 88.2 percent.

Non-metropolitan Changes

The Midwest has grown slowly because of economic adjustments in both rural and urban areas. Urban areas have fared better, however, as new forms of industrial and service production have replaced part of the large industrial and agribusiness base. Accordingly, the population growth that did occur in the Midwest during the 1980s was concentrated largely in the metropolitan areas. In aggregate, the region's 194 metropolitan counties added 1.1 million (2.6 percent) to their population, while the region's 856 non-metropolitan counties collectively lost 1.7 percent. Seventy-one percent of the metropolitan counties gained population, compared with only 28 percent of the non-metro counties. The metropolitan centers, however, exert an economic and demographic influence on nearby rural counties. Of the 856 non-metro counties in the region, 293 lie next to a metropolitan area; 563 are located more remotely. Forty-two percent of the adjacent rural counties gained

TABLE 2.1 U.S. Population and Change by Region, 1960-1990 (in thousands)

	1960	1970	Percent Change 1960-1970	1980	Percent Change 1970-1980	1990	Percent Change 1980-1990	Percent Change 1960-1990
U.S.	179,323	203,302	13.4	226,546	11.4	248,710	9.8	38.7
Northeast	44,678	49,061	9.8	49,135	0.0	50,809	3.4	13.7
Midwest	51,619	56,589	9.6	58,866	4.0	59,669	1.4	15.6
South	54,973	62,812	14.2	75,372	20.0	85,446	13.3	55.4
West	28,053	34,838	24.2	43,172	23.0	52,786	22.2	88.2

Source: United States Department of Commerce, Population Census.

population during the 1980s, compared with only 21 percent of the more remote rural counties.

In the 1980s, rural economic conditions deteriorated more severely in the Midwest than in other regions. Table 2.2 compares income, employment, and population changes in each region's non-metro counties. Non-metro population declined only in the Midwest, and the Midwest also experienced the greatest loss in farm population. In each of the other regions the non-metro population increased, but less than in metropolitan areas. The non-metro Midwest also had the lowest rate of increase in wage and salary employment and per capita income during the 1980s. In addition, it suffered the second largest increase in percentage of population below the poverty line, although the poverty rate in the rural Midwest remains well below rates in the rural South and West.

TABLE 2.2 Non-metropolitan Economic and Demographic Changes Among Regions, 1980-1990

	Percent Change in Population	Percent Change in Farm Population	Percent Change in per Capita Income	Percent Change in Wage and Salary Employment	Percent Population Below Poverty Income 1990	Percent Change in Population Below Poverty 1980-1990
Northeast	3.9	-30.2	59.6	16.2	11.9	2.3
Midwest	-1.7	-34.0	31.7	9.6	13.5	10.3
South	4.6	-33.7	49.8	12.7	20.6	7.6
West	14.6	-23.5	40.2	17.4	16.2	34.2

Source: United States Department of Commerce, Bureau of Economic Analysis and 1990 Population Census.

Variations Within the Midwest

Although the Midwest is clearly different from other regions, equally dramatic differences exist within the region. Of the 1,050 counties of the Midwest, 673 (64 percent) lost population, while 274 gained some population. Only 108 gained 10 percent or more.

Most rural counties in Iowa, Nebraska, the Dakotas, Kansas, downstate Illinois, northern Missouri, and western Minnesota lost population. Generally the declining counties are located beyond commuting distance to larger metropolitan labor markets, and their economic base depends heavily on family farm agriculture. Those counties continue to lose population largely because of the outmigration of youths and younger adults. Their migration leaves behind a higher proportion of older persons. As a result, Iowa, Missouri, Kansas, Nebraska, and the Dakotas rank high nationally in the proportion of the population age 65 and over. The elderly of those states disproportionately populate the rural areas and small towns.

The counties whose population gained by more than 10 percent during the 1980s generally fit one of three descriptions:

1. They surround large cities. For example, the gainers include nine counties surrounding Minneapolis-St. Paul, eight counties surrounding St. Louis, and five counties surrounding Chicago. The absence of larger cities and labor markets is a major factor contributing to the more widespread population loss in Iowa, the Dakotas, Nebraska, and Kansas. Major parts of Ohio, Indiana, and Illinois and the lower halves of Michigan and Wisconsin are composed of counties that either are classified as metropolitan or are adjacent to metropolitan counties. Such locations increase the prospects that farm operators and their families will find off-farm employment to supplement their farm income.
2. A number of rural counties in northern Michigan, Wisconsin, and Minnesota, in the Black Hills of South Dakota, and in southern Missouri have become attractive to retirees relocating from other parts of the country. The topography and climate of those locations limit their agricultural productivity but offer recreational amenities and low property values, which are attractive to retirees.
3. Some of the smaller cities and metropolitan areas are home to some of the region's major public universities. This category includes counties containing the University of Nebraska, the University of Missouri, the University of Wisconsin, the

University of Iowa, Michigan State University, the University of Indiana, and others.

Distribution and Sources of Income

The context of agriculture and of rural residents of the region also is reflected in the distribution and sources of income. The region contains differences not only in population change but also in per capita income. Differences in income can be explained in part by some of the trends and geographic differences discussed above. In general, rural counties in the region with the highest per capita income tend to fit one of two descriptions:

1. Counties within commuting range of large metropolitan areas tend to enjoy higher per capita income. Such counties are concentrated near Chicago, Milwaukee, Minneapolis-St. Paul, Detroit, Indianapolis, Cleveland, Columbus, Cincinnati, St. Louis, Kansas City, and other cities.
2. Numerous rural/agricultural counties scattered across Kansas, Nebraska, Minnesota, Iowa, and Illinois also have high per capita income. These counties are dominated by larger-volume commercial farms but have low population density. This is the case especially in western Kansas and Nebraska, where large-scale irrigated crop production and cattle feedlots are concentrated. In such counties, per capita income is high both because income has been increasing and because population has been declining. With respect to production agriculture, increase in per capita income does not necessarily improve prospects for population growth.

Low-income counties in the region tend to be concentrated in areas that have relatively low agricultural productivity, are outside commuting range to metropolitan areas, and/or recently have attracted relocating retirees. An exception is the prevalence of lower-income counties in western North and South Dakota. A number of these counties, as well as some counties in northern Minnesota and Wisconsin, are the sites of Native American reservations.

We have mentioned the growing proportion of older persons in many parts of the region, both because of the continued outmigration of young adults and because some regions are attractive to relocating retirees. As a result, a growing proportion of income is derived from transfer payments (basically government entitlement programs exclusive of agricultural loans and payments). Two categories of people are most likely to receive transfer payments: the retired (e.g., Social Security, Medicare) and low-income persons (e.g., Aid to Families with Dependent Children, food stamps). The counties in which more than 20 percent of total income is derived from transfer payments generally are those in which the highest proportions of retired persons live. Many such persons have retired from non-farm employment and have moved to small farms for a combination of lifestyle and economic reasons.

Because of the farm crisis of the 1980s and the related loss in rural businesses, a growing number of rural families have incomes below the federal poverty guideline. In rural counties scattered throughout the region, 20 percent or more of young families with children have incomes below poverty level. Many are family farm agricultural counties in which numerous families lost their farms.

Family Farm Agriculture in the Midwest

Although the number of farms in the region continues to decline (from 932,000 in 1982 to 862,000 in 1987) the farming character remains, even in the more industrial eastern states. The 12 North Central states account for 24 percent of the 1990 U.S. population and 22 percent of the nation's geographic area, but those states together include 41 percent of the nation's farms, which combined to produce 42 percent of the nation's 1987 agricultural output. In no other region are farms so numerous and collectively so productive. Also, as we will describe below, the Midwest is the region in which the most farm families depend exclusively on their farms for their livelihood. Accordingly the burden of the early 1980s farm crisis was not shared equally by all rural regions of the country; the effects were borne disproportionately by farm families of the Midwest, especially in the western corn belt and the plains states.

Although the Midwest includes the nation's farming heartland, farming varies greatly within the region. The corn belt indeed is a belt, stretching from Ohio halfway across Kansas, Nebraska, and South Dakota. Among those states, however, only Iowa can be described as located wholly within the corn belt. Residents of all other states generally regard their state as divided into at least two distinctive regions, separated according to topography, rainfall, and/or productivity of agriculture. Michigan, Wisconsin, and Minnesota usually are divided into north and south, with the southern part included in the corn belt. The most productive agriculture in those three states is to be found in the south. The northern parts are extensively wooded and are attractive for outdoor recreation, retirement, and a variety of economic activities including mining. (Subsequent analysis in this volume refers to these three as the lakes states.) Ohio, Indiana, Illinois, and Missouri also are divided into north and south, but in those states the northern part is included in the corn belt and the southern part consists of more rugged and heavily wooded terrain. Kansas, Nebraska, and the Dakotas typically are divided into east and west; the eastern portion is included in the corn belt and the west generally is included in the more arid great plains. (In later analysis these four are grouped as the plains states.) Western Kansas and Nebraska, however, now are characterized by highly productive large-scale irrigated agriculture, cattle feedlots, and associated meat processing. The non-irrigated areas of western Kansas and Nebraska, like the Dakotas are devoted extensively to cattle ranching. Because cattle ranchers also are family farmers, they generally did not fare any better economically during the past decade than their corn belt counterparts.

The regional variation among the states is reflected further in the number, size, and type of farms to be found in each. As shown in Table 2.3, Missouri leads the region with 106,000 farms, followed closely by

TABLE 2.3 Number of Farms by Gross Farm Commodity Sales Categories, 1987

Area	Number of Farms	$500,000 or More	$100,000-$499,999	$50,000-$99,999	$10,000-$49,999	Under $10,000
			------sales category------			
NC Region	861,982	1.1	16.2	15.2	32.0	35.5
Illinois	88,786	1.2	20.9	16.6	31.8	29.4
Indiana	70,506	1.2	14.4	11.4	30.5	42.6
Iowa	105,180	1.5	23.9	19.6	33.5	21.4
Kansas	68,579	1.4	12.3	13.1	35.5	37.7
Michigan	51,172	1.1	11.3	8.4	26.3	52.7
Minnesota	85,079	1.0	18.3	18.1	31.2	31.4
Missouri	106,105	0.5	7.9	8.4	30.1	53.1
Nebraska	60,502	2.1	21.0	18.7	35.7	22.5
N. Dakota	35,289	0.7	16.2	22.1	41.2	19.8
Ohio	79,277	0.7	10.1	9.7	29.8	49.7
S. Dakota	36,376	1.2	17.4	21.2	37.9	22.3
Wisconsin	75,131	0.8	19.6	20.3	28.2	31.1
U.S.	2,087,759	1.5	12.6	10.4	26.1	49.2

Source: United States Department of Commerce, Bureau of the Census, 1987 Census of Agriculture.

Iowa with 105,000 (Missouri ranks second nationally behind Texas in the number of farms; Iowa ranks third). Although Missouri and Iowa have about the same number of farms, the similarity ends there. In 1987, total cash commodity sales for Iowa farms fell just short of $9 billion; total farm commodity sales in Missouri were only $3.6 billion (Table 2.4). The fewest farms are found in North Dakota and South Dakota, with 35,000 and 36,000 farms respectively.

Table 2.3 not only shows the total number of farms per state but also classifies them according to farm size, measured by total cash agricultural commodity sales. As farms have become more specialized and more commercial, cash sales of commodities have been used as a measure of farm size more often than the number of acres farmed. Some kinds of "farms," such as cattle feedlots and confinement swine operations, require very little land but generate a large dollar sales volume. Therefore cash sales reflect the size of a farm business more accurately than acreage.

The U.S. Agricultural Census, from which these data are obtained, counts a business as a farm if it sells at least $1,000 of agricultural

TABLE 2.4 Percentage of Total Agricultural Products Sold by Each Gross Sales Category of Farms, 1987, by State

Area	Total Sales ($ Thousands)	$500,000 or More	$100,000-$499,999	$50,000-$99,999	$10,000-$49,999	Under $10,000
NC Region	57,633,116	24.6	45.0	16.3	12.1	2.0
Illinois	6,376,801	15.3	54.9	16.8	11.5	1.6
Indiana	4,067,684	21.6	48.8	14.1	12.7	2.7
Iowa	8,926,799	18.4	53.4	16.7	10.5	1.0
Kansas	6,476,669	54.6	24.8	9.8	9.2	1.6
Michigan	2,545,078	26.2	45.8	12.2	12.3	3.6
Minnesota	5,676,376	16.8	49.7	19.5	12.3	1.7
Missouri	3,644,988	14.0	42.9	17.5	19.9	5.7
Nebraska	6,667,132	42.4	35.9	12.2	8.5	0.9
N. Dakota	2,188,158	9.7	45.4	25.3	18.1	1.4
Ohio	3,434,064	20.1	43.7	15.9	16.2	4.1
S. Dakota	2,719,498	23.5	41.4	20.3	13.6	1.2
Wisconsin	4,909,869	13.1	51.1	22.6	11.4	1.8
U.S.	136,048,516	38.2	38.2	11.5	9.6	2.5

Source: United States Department of Commerce, Bureau of the Census, 1987 Census of Agriculture.

produce annually. In the Midwest, however, that figure scarcely qualifies a household as a farm. On a productive farm in a good year, operators hardly ever realize a net income of as much as 20 percent of their gross sales. Thus, under the best of conditions, gross farm sales of $100,000 seldom would yield as much as $20,000 net income. During the farm crisis, net farm income was often negative, even for larger farms.

In the North Central region, 1 percent of the farms were classified as selling more than $500,000 worth of crops and livestock in 1987; 16 percent sold between $100,000 and $499,000; 15 percent sold between $50,000 and $99,000; and 68 percent sold less than $50,000.

Farms in the $50,000-$99,000 sales range are often described as "small family farms," those in the $100,000-$500,000 range as "family farms," and those in the over-$500,000 category as "larger-than-family farms" or sometimes as "industrial" farms. Compared with other regions, the North Central region has the highest proportion of farms in the $50,000-$500,000 (family farm) categories.

Farms in the under-$50,000 category represent a recent and growing phenomenon in agriculture, both nationally and in the region. Such "farms" frequently are called "part-time" because their operators usually depend primarily on non-farm sources of income for their livelihood. It is significant that more than two-thirds of the "farms" in the Midwest belong to this category. As the following chapters will show, the crisis of the 1980s increased the number of such farms as many farm families adapted to economic distress in farming by reducing their farming operation and taking full-time employment off the farm.

Nebraska leads the region in the proportion of larger-than-family farms (2.1 percent); Iowa has the highest proportion of family farms (23.9 percent). Missouri leads in the proportion of part-time farms (83.2 percent), followed closely by Ohio (79.5 percent) and Michigan (79.0 percent). Half or more of the farms in those three states sell less than $10,000 worth of agricultural commodities per year. These percentages are large because many of the farms are close to larger urban labor markets (which create more off-farm employment opportunities) and because of the relatively large proportion of rough and forested land in those states.

Proportion of Production Generated by Farms in Different Size Classes

Although farms with sales of more than $100,000 make up only 17 percent of the region's total farms, they generate 69.6 percent of the region's total agricultural output (Table 2.3 and 2.4). In Kansas 54.6

percent of total agricultural commodity sales is generated by farms selling more than $500,000; in Nebraska 42.4 percent of total production is generated by the largest class of farms. In the remaining ten states, production is concentrated more in the family farm ($100,000-$500,000) category. In Iowa, Illinois, Wisconsin, and Minnesota more than 50 percent of the state's total sales is produced by farms in that range. The degree of concentration of production in the top two sales categories is substantial. Kansas leads the region; 79.4 percent of its farm production is generated by farms with sales of more than $100,000. Nebraska follows with 78.3 percent of statewide production. Between 70 and 72 percent of total gross sales is produced by family-size and larger farms in Iowa, Illinois, Michigan, and Indiana. The lowest proportions of state output produced by family-size and larger farms are found in Missouri (56.9 percent) and North Dakota (55.1 percent).

The Midwest includes 90 of the nation's top 200 agricultural producing counties (as measured by total cash value of farm commodities sold) but it contains only 11 of the nation's top 50 agricultural producing counties. Seven of those 11 are concentrated in southwestern Kansas and reflect the large-scale cattle feedlots and irrigated farming in that region. Most of the remaining top 50 counties are dominated by larger-than-family farms and are located in California, Texas, and Florida.

Beyond the top 50, however, the Midwest includes 79 of the nation's remaining top 200 agriculture producing counties. Iowa leads the region with 22 of the 200.

Net Cash Returns from Farming

Although family farms typify the North Central region, farms in the 12 states vary substantially because of the relative size and location of the farms. For the region and for each state, Table 2.5 shows the contribution of different sizes of farms to each state's total net cash returns from farming. For 1987 the 12 states combined to generate 46 percent of the nation's net cash returns from agriculture. Iowa led the region in total cash returns ($2.15 billion), followed by Illinois ($1.73 billion). For the region, farms producing sales of more than $500,000 contributed 18.5 percent of the region's net cash returns from farming, although 1.1 percent of the region's farms belonged to this category. Farms in this size category generated 24.6 percent of the region's cash commodity sales. The 16 percent of farms with sales between $100,000 and $500,000 generated 45 percent of the region's total cash sales but 56 percent of net cash returns. The 15 percent of farms with sales between

TABLE 2.5 Farm Net Cash Returns by State and Size of Farm, 1987

Area	Total Net Cash Returns ($ Thousands)	$500,000 or More	$100,000-$499,999	$50,000-$99,999	$10,000-$49,999	Under $10,000
			----Sales Category----			
NC Region	12,141,621	2,246,467	6,834,498	2,398,647	1,161,569	(499,564)
Illinois	1,730,879	236,064	1,075,888	310,518	143,958	(35,550)
Indiana	883,642	173,139	514,185	146,503	89,687	(39,871)
Iowa	2,146,997	305,240	1,310,877	396,972	174,432	(40,526)
Kansas	922,225	384,085	357,158	132,452	82,573	(34,043)
Michigan	319,953	105,754	231,452	50,592	11,566	(79,411)
Minnesota	1,233,896	171,082	713,744	284,585	115,903	(51,417)
Missouri	829,853	102,327	444,303	174,413	159,400	(50,590)
Nebraska	1,229,040	333,063	614,106	196,939	103,994	(19,063)
N. Dakota	332,673	39,091	195,894	84,411	28,043	(14,765)
Ohio	685,358	136,764	368,261	136,571	103,467	(59,706)
S. Dakota	573,997	106,149	291,720	131,793	58,010	(13,675)
Wisconsin	1,253,108	153,709	716,910	352,898	90,536	(60,947)
U.S.	26,430,535	9,546,065	12,715,261	3,829,699	2,108,536	(1,769,026)

Source: United States Department of Commerce, Bureau of the Census, 1987 Census of Agriculture.

50,000 and $100,000 produced 16 percent of the region's cash sales and 20 percent of net cash returns.

Profits from farming (in terms of sales) are poorest for the smallest farms. Those farms, although 68 percent of the region's total farms, generate only 14 percent of cash sales and only 5 percent of net cash returns. The aggregate of net cash returns from farms with sales of less than $10,000 is negative both for the region and for all 12 states.

The "farms" with sales of less than $10,000 deserve specific mention. They are counted as farms by the U.S. Census of Agriculture, which applies its definition of a farm to any enterprise that sells at least $1,000 worth of farm commodities per year. Although they are farms by definition, the 35 percent of the North Central region farms that sell less than $10,000 worth of farm commodities do not represent the major source of income for most of their operators. Analysts recently have begun to refer to such "farms" as "rural residences" or even as "hobby" farms. The operator frequently is retired from some other occupation, or is employed full-time off the farm. For many such farms, the operator does not depend on the farm for income, but instead subsidizes it from non-farm earnings. Because such rural residences frequently are operated in conjunction with retirement or with other full-time off-farm work, they tend to be concentrated in the states and regions that offer those opportunities. Accordingly, the highest percentages of such farms are found in southern Missouri, northern Michigan, Indiana, and Ohio, where many urban labor markets are distributed uniformly.

The growing prominence of part-time farms is reflected in farmers' reports of their principal occupation. Across the region, 63 percent of the "farmers" report farming as their principal occupation, in comparison with 55 percent for the nation. The proportion of farmers who claim farming as their occupation varies from a low of 50 percent in Ohio and Missouri to a high of 82 percent in North Dakota. In Iowa, Nebraska, South Dakota, and Wisconsin, more than 70 percent regard farming as their principal occupation. Clearly, claiming status as a farmer is associated strongly with the cash sales classification of the farm: for the region, 92 percent of farmers who produce between $100,000 and $500,000 in cash sales report farming as their principal occupation. Interestingly, a somewhat smaller percentage (90 percent) of those who produce more than $500,000 claim to be full-time farmers. Only about one-third of operators of farms with cash sales of less than $10,000 claim farming as their occupation. It is reasonable to speculate (on the basis of the high average age of those operators) that a substantial proportion of the 32 percent in the last category depends principally on some form of retirement income.

Commodity Production in the North Central Region

Although Midwestern agriculture produces a variety of crops and livestock, a few commodities historically have been, and continue to be, dominant in the region.

The "corn belt" remains an appropriate name for much of the region; the 12 states produce 87 percent of the nation's total corn crop (see Table 2.6). In addition to corn, the North Central region adapted well to soybean production when that crop was introduced commercially about 50 years ago. Soybeans proved to be complementary with corn production and have become a part of a rotation with corn on many of the region's cash grain farms. Accordingly the North Central region has become as nationally dominant in soybean as in corn production; in 1987 it produced 82 percent of the nation's soybean crop.

In addition to corn and soybeans, the region historically has been a major contributor to the nation's wheat production. As mentioned earlier, the four western plains states can be divided into east and west largely on the basis of topography and rainfall. The eastern half of

TABLE 2.6 Commodity Production, 1987

Commodity	North Central Production as a Percentage of U.S. Production	North Central Region: Top Three States and Their Percentage of the Region's Production
Corn (bu)	86.9	Iowa (21.8) Illinois (20.8) Nebraska (12.8)
Soybeans (bu)	82.1	Illinois (21.8) Iowa (21.6) Indiana (11.2)
Wheat (bu)	52.1	Kansas (29.8) North Dakota (25.3) Minnesota (10.0)
Hogs and Pigs (number sold)	79.4	Iowa (30.6) Illinois (12.9) Indiana, Minnesota (10.5)
Cattle/Calves (dollar value)	45.5	Kansas (27.4) Nebraska (22.8) Iowa (13.5)
Cattle/Calves (number sold)	39.6	Nebraska (15.4) Kansas (14.6) Iowa (11.3)
Dairy Products	41.8	Wisconsin (41.8) Minnesota (15.8) Michigan (8.5)

Source: United States Department of Commerce, Bureau of the Census, 1987 Census of Agriculture.

each state generally is included in the corn belt; the western half traditionally is devoted to dryland wheat farming and cattle ranching. Wheat remains a dominant crop in the plains states; in 1987 the region accounted for 52 percent of national wheat production.

Because corn is used largely for red meat production and because soybean oil meal is a widely used protein supplement, the region concomitantly has been historically dominant in swine and beef production. The North Central region accounts for 79 percent of the nation's swine production and 46 percent of the cash value of cattle and calves sold. These concentrations remain in the region even though production of both commodities recently has tended to move toward the South and the West.

Dairying also remains an important sector of the region's agricultural economy. In 1987 the region produced 42 percent of the nation's dairy output, although only two states—Wisconsin and Minnesota—accounted for 58 percent of the region's dairy production.

A pattern of regional specialization exists in commodity production, as shown by the proportion of each commodity produced by the leading producer states. In the case of corn, the leading three states (Iowa, Illinois, Nebraska) account for 55 percent of the region's production; three states (Illinois, Iowa, Indiana) also account for 55 percent of soybean production; for wheat, Kansas, North Dakota, and Minnesota produce 65 percent of the regional total; for hogs, Iowa, Illinois, and Indiana produce 54 percent; for cattle and calves, Kansas, Nebraska, and Iowa produce 65 percent; for dairy products, Wisconsin, Minnesota, and Michigan produce 66 percent.

Although the North Central region continues to dominate national production of these commodities, the place and manner of production have changed substantially. North Central farms have become increasingly specialized in producing either cash grain or livestock and livestock products. As recently as 40 years ago, most farms in the region produced each of the commodities discussed above. Those "general" farms raised grain and several species of livestock, and frequently produced dairy commodities as well. The general farms of that earlier era tended to manage risk through diversity of production. As farms have become more specialized, they have become concomitantly more efficient but also more vulnerable to risks affecting their commodity production. Government farm programs of the past two decades have contributed to spreading some of that risk and have become important to the income of the region's farmers. Government programs apply to four of the major commodities produced in the region: corn, soybeans, wheat, and dairy products.

Conclusion

While the Midwest appropriately continues to declare itself the nation's farm belt, economic, political, and technological changes are continuing to redefine farms, who is operating them, and why. The Agricultural Census reports 862,000 farms in the 12-state region, but the same census reveals that only 149,000 farms (the 17 percent that sell more than $100,000 worth of agricultural commodities) are producing 70 percent of the region's total farm output. When farms with sales between $50,000 and $100,000 are added, the combined total of 280,000 farms (32 percent) is producing 86 percent of the region's output.

These data show clearly that two different agricultures coexist in the traditional farm belt. One agriculture (the 32 percent producing at least $50,000 in sales) contributes most of the region's commodity output. That agriculture continues to be dominated by an updated version of family farms, although the data make clear that in some parts of the region and for some commodities, larger-than-family farms are contributing an increasing proportion of the total.

Yet the presence of 582,000 "farms" (68 percent) in the region that use other sources of off-farm income to maintain the farm suggests the continued importance of family farm values and lifestyles. For the smallest one-third of the farms, as stated above, evidence shows that operators' off-farm income is being used in fact to subsidize the operating losses of a farm. The continued existence of these "rural residence" farms depends heavily on their operators' continued access to off-farm employment and income-generating possibilities. The farm and the non-farm sectors of the rural Midwest remain closely linked, although this link differs greatly from the farm-community interdependence of earlier times.

References

Flora, Cornelia Butler. 1990. "Presidential Address: Rural Peoples in a Global Economy." *Rural Sociology* 55(2): 157-177.

Schotsch, Linda and Dick Seim. 1980. "Can You Make $3,000 Land Pay?" *Farm Journal* 104 (February): 11-13.

Wennblom, Ralph. 1980. "Bergland Sees No Farm Crisis Yet." *Farm Journal* 104 (June/July): 44-45.

PART ONE

The Farm Enterprise

Introduction
F. Larry Leistritz

As noted in the introductory chapter, this book has three foci. These are: (1) the farm enterprise, (2) the farm household, and (3) the rural community. This section of the book focuses on the farm enterprise. Specifically, the three chapters that follow provide a view of the financial characteristics of farms across the North Central region, the adjustments and adaptations their operators made in coping with the conditions that existed during the period 1984-1988, and the operators' plans for change during the next five years. In addition, the operators' views regarding their needs for information and training are examined.

Chapter 3 provides a snapshot of the financial status of farms and farm households in the North Central region at the time of the survey. Gross farm sales offer one means of comparing the scale of the farming operation across various parts of the region and among enterprises. Net family income, which includes income from both farm and non-farm sources, provides a measure of the financial well-being of farm households while the composition of that income (farm income, off-farm earnings, and other non-farm income) gives an indication of the dependence of various types of farm households on farm income compared to income from other sources. Balance sheet information (debt, assets, and net worth) also is examined to gain insights regarding the financial strength or solvency of farming operations. Additional insights concerning the financial status of farms in the North Central region are gained by examining delinquency rates and farmers' expectations regarding future financing problems. These financial measures are analyzed to identify differences that may exist among subregions and among farms of different sizes.

Chapter 4 focuses on adaptations that farm operators had made in coping with economic conditions during the five years preceding the survey. The types of adaptations and adjustments examined include (1) those which increase income from farming or from non-farm sources or which improve efficiency, (2) those which reduce costs or postpone expenditures, (3) those which transfer resources out of farming, and (4) other actions that minimize exposure to risk. The patterns of operators' adaptations are examined by subregion, by farm size, and by the age of the operator.

Chapter 5 examines changes that farm operators plan to make within the next five years. The types of changes examined are similar to those discussed in Chapter 4 and include (1) diversification of farm enterprises, (2) changing control over land (rent or own more or fewer acres), (3) changes in the owner's involvement in farming, (4) reduction in risk exposure, (5) reduction in cash expenditures, and (6) financial management and marketing. Farm operators also provided information about the types of information and training they believed they would need in coming years. Areas of information and training examined include (1) production technology, (2) business management and marketing, (3) new inputs and enterprises, and (4) on-farm processing. The patterns of planned adaptations and information and training needs are examined by subregion, farm size, and operator age.

Together these chapters provide important insights regarding the current status and likely future changes in the farms of the North Central region. This information in turn should be helpful in understanding household and community changes and adaptations.

3

Financial Characteristics of Farm Operators

F. Larry Leistritz and Freddie L. Barnard

In the decade preceding the regional farm survey, many areas of rural America experienced a drastic reversal of fortunes. From a period of unparalleled growth in asset values and farm income and widespread optimism concerning the future, the farm sector was plunged into the most severe economic stress, in many respects, since the 1930s (Leistritz and Murdock 1988). Between 1981 and 1986-1987, land values fell sharply (in several North Central states by more than 50 percent) and net farm income (adjusted for inflation) receded to only a fraction of the levels attained in the mid-1970s. Farm foreclosures, forfeitures on land contracts, and defaults on notes became widespread. Even farm families that were able to meet their payments and retain their assets saw their net worth shrink dramatically. By spring 1989, however, when the North Central Regional Farm Survey was conducted, the worst appeared to be past for producers in much of the region, although drought and depressed grain markets slowed recovery in some areas while strong livestock prices provided a stimulus in others.

Widespread concern about economic conditions in agriculture during the 1980s led to a number of studies of farm financial conditions (Dobson, Barnard, and Graves 1985; Doye, Jolly, and Choat 1987; Johnson, Morehart, and Erickson 1987; Leholm, Leistritz, Ekstrom, and Vreugdenhil 1985). These studies used various measures of the solvency and liquidity of the farm business or household, as well as measures of the profitability of the farm business (Leistritz and Ekstrom 1988).

This research, however, was limited by the fact that 1) the studies which were made on a national basis generally reported only a limited amount of information regarding the characteristics of the farm business and farm household, thus restricting the ability to assess relationships between these attributes and measures of financial stress, and 2) the studies that provided more information about family characteristics, farm management practices, and household adjustments to financial stress typically were limited to single states or areas of states.

Although only limited generalizations can be drawn from previous studies, a few conclusions appear particularly relevant to this discussion. First, when measures of farm financial vulnerability have been compared among regions, the North Central states generally have been found to have the largest percentages of their farms in the highly vulnerable categories. Second, the percentage of farms with both high debt-asset ratios and negative cash flow was generally greatest among farms with gross sales between $40,000 and $250,000, the sales classes that many would describe as family-size commercial farms (Leistritz and Ekstrom 1988). These findings show that family farms in the North Central states were hit particularly hard by the farm financial crisis of the 1980s.

The objective of this chapter is to provide information about the financial characteristics of farm operations in the North Central region. We draw inferences from survey information relating to gross farm sales and net family income, off-farm employment and other non-farm income, debt-asset ratios, delinquency rates for loans, loan rejection rates, and other indicators.

The diversity in farming in the 12-state North Central region makes comparisons among the 12 states difficult. For more meaningful comparisons, the regional sample was divided into three subregions: the corn belt (Illinois, Indiana, Iowa, Missouri, and Ohio); the plains states (North Dakota, South Dakota, Kansas, and Nebraska); and the lakes states (Minnesota, Michigan, and Wisconsin) (see Figure 3.1).

Questions regarding financial condition were included only on the survey mailed to operators. The weighted sample size of the corn belt subregion is 2,112 operator respondents. The plains subregion has a weighted sample size of 947 operators, the lakes subregion has a weighted sample size of 1,028 operators.

Gross Farm Sales and Net Family Income

Gross farm sales is the income generated by a farm from the sale of farm products (including government farm payments) before expenses

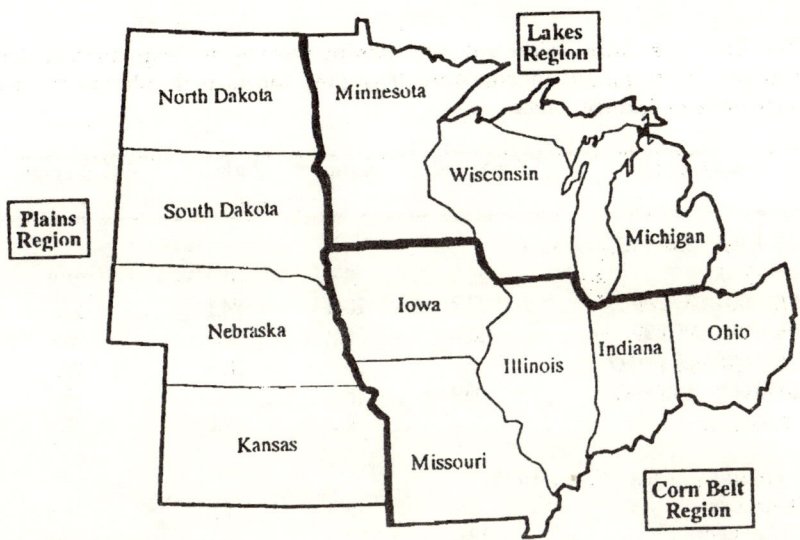

FIGURE 3.1 North Central Region and Subregions

are subtracted, to be used in calculating the different measures of net farm income. The cutoff point that sometimes is used for differentiating between part-time and full-time farms is $100,000 of gross farm sales. This figure is arbitrary and does not apply to every farm operation when part-time farms are distinguished from full-time farms. If the $100,000 cutoff point is used, about 30 percent of the respondents in the region would have qualified as operators of full-time farms and 70 percent as operators of part-time farms in 1988.

The comparable figures for the three subregions regarding full-time and part-time farmers are reported in Table 3.1. The figures show that the percentage of respondents with gross farm sales of less than $100,000 in the corn belt (73.5 percent) is similar to the percentage in the lakes subregion (71.4 percent). The percentage of respondents in this category is lower in the plains subregion (61.1 percent).

These figures show that many operators of smaller, part-time farms in the region obtain large percentages of their income from off-farm employment and other non-farm sources. Because the financial problems on part-time farms may differ from those on full-time farms, we present certain statistics separately for farmers with gross farm sales of less than $100,000 per year and for farmers with gross farm sales greater than this amount.

TABLE 3.1 Gross Farm Sales and Net Family Income of Respondents, and Percentage Total Family Income from Farming, Off-farm Employment, and Other Non-farm Sources, 1988

Category	Corn Belt	Plains	Lakes	NC Region
	----------percent----------			
*Gross Farm Sales Categories**				
< $10,000	22.5	9.5	21.3	19.1
$10,000 - $39,999	27.7	20.6	24.1	25.2
$40,000 - $99,999	23.3	31.0	26.0	25.8
$100,000 - $249,999	19.7	28.1	21.9	22.2
$250,000 - $499,999	4.9	8.5	4.7	5.7
≥ $500,000	1.9	2.3	2.0	2.0
*Net Family Income Categories**				
Net Loss	5.1	5.8	5.9	5.4
$1 - $9,999	12.0	12.2	18.1	13.6
$10,000 - $19,999	22.5	22.8	23.9	22.9
$20,000 - $29,999	19.9	22.4	21.6	20.9
$30,000 - $39,999	17.4	14.0	12.8	15.4
$40,000 - $49,999	10.0	9.4	7.6	9.2
$50,000 - $59,999	5.3	5.1	4.3	5.0
$60,000 - $69,999	3.5	3.3	2.4	3.2
≥ $70,000	4.4	5.1	3.4	4.3
Total Family Income Categories (all respondents)				
Farming	52.5	73.5	58.9	59.2
Off-farm employment	34.2	18.4	29.0	29.0
Other non-farm income	13.3	8.1	12.1	11.8
Total Family Income Categories (gross farm sales < $100,000)				
Farming	42.1	62.7	49.1	48.3
Off-farm employment	41.7	25.9	36.1	36.9
Other non-farm income	16.2	11.4	14.8	14.8
Total Family Income Categories (gross farm sales ≥ $100,000)				
Farming	80.8	88.1	83.8	83.9
Off-farm employment	14.6	8.4	11.4	11.8
Other non-farm income	4.6	3.5	4.8	4.3

* Gross farm sales categories were defined to include government farm payments. Net family income includes income from off-farm employment, farming, and other non-farm sources (e.g., interest and Social Security).

Respondents also reported net family income for 1988. For purposes of the survey, net family income includes income from off-farm employment, farming, and other non-farm sources such as interest and Social Security. Net family income is defined as the return from farming to unpaid operator and family labor, management, and equity capital plus income from off-farm employment and other non-farm sources. Viewed another way, net family income and depreciation allowances represent the amount of money available to farmers to repay the principal on concurrent liabilities (intermediate and long-term debt), to purchase capital assets, to pay family living expenses, to pay income taxes, and to retain in the farming operation as a financial reserve.

About 42 percent of the respondents had net family incomes of less than $20,000 in 1988. In view of recent levels of family living expenses, many of these farmers probably would have been unable to make debt payments from their 1988 net family incomes. The percentage of families with net family incomes of less than $20,000 was highest in the lakes subregion (47.9 percent), followed by the plains (40.8 percent), and the corn belt (39.6 percent).

Many farmers in the region supplement farm income with income from off-farm employment and other non-farm sources, which can reduce the problems created by low net farm incomes. Respondents were asked to report the percentage of total family income for 1988 derived from farming (including government farm payments), off-farm employment (including operator and spouse), and other non-farm sources (e.g., interest, Social Security). Regionwide the respondents reported that about 59 percent of their total family income came from farming, 29 percent from off-farm work, and about 12 percent from other non-farm sources (Figure 3.2). The percentages for the lakes subregion were similar to those for the whole region; the corn belt respondents reported slightly more of their income (34.2 percent) from off-farm employment and less from farming (52.5 percent). The plains subregion respondents, however, reported a much higher percentage of their income from farming (73.5 percent), while their income from off-farm employment and from other off-farm sources was substantially less.

The percentage of income from off-farm employment for farmers with less than $100,000 of gross farm sales exceeded that for the larger farmers. This pattern probably reflects the heavy demands placed on the operators of larger farms for farm labor, the smaller number of cases on larger farms in which both the farmer and the spouse work off the farm, the incomes generated on the larger farms, which lessen the need for off-farm work, and other factors.

Regionwide, operators of farms with less than $100,000 gross sales

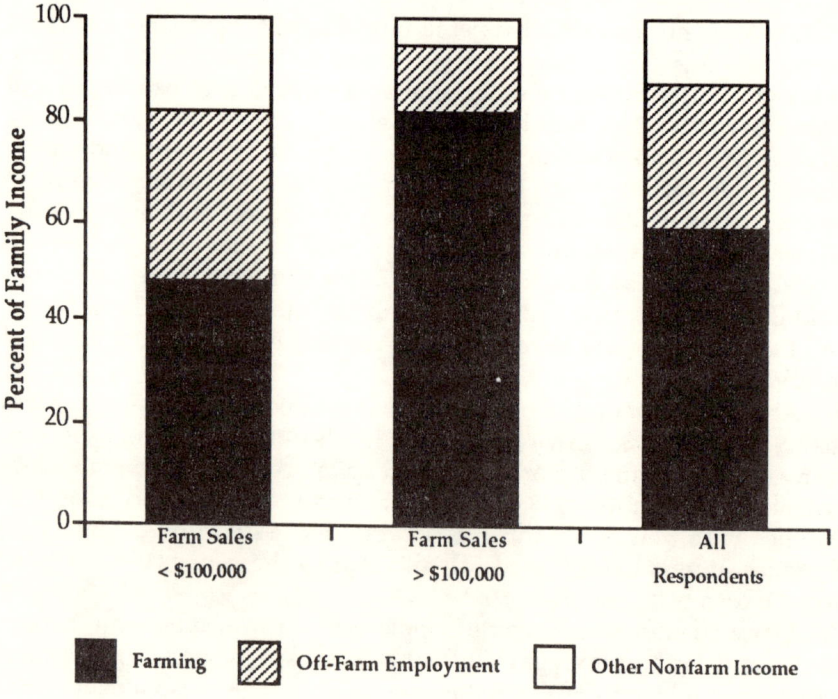

FIGURE 3.2 Sources of family income for respondents with gross farm sales less than $100,000, sales of $100,000 or more, and all respondents, North Central Region, 1988

relied on the farming operation for only 48 percent of their total family income. This proportion ranged from a low of 42 percent in the corn belt to a high of almost 63 percent in the plains. Corn belt operators of these smaller farms obtained almost as much income from off-farm employment of household members (41.7 percent) as from farming (42.1 percent). Operators of smaller farms in the plains, however, rely more heavily on farming (62.7 percent) as a source of family income than in the other two subregions.

On the other hand, farmers regionwide with gross sales over $100,000 obtained 83.9 percent of total family income from farming. The proportion was greater than 80 percent in all three subregions. The highest percentage was in the plains subregion (88.1 percent); the lowest percentage was in the corn belt (80.8 percent).

The figures quoted above illustrate the similarities between the corn belt and the lakes subregions regarding the percentage of respondents with less than $100,000 in gross farm sales and the importance of off-farm employment as a source of family income for those respondents. The figures also illustrate the importance of farming as a source of family income for all respondents in the plains subregion, even those with less than $100,000 in gross sales.

Balance Sheet Information Used to Obtain Measures of Solvency

A balance sheet is a financial picture of an individual or firm at a given point in time, which shows assets (what is owned), liabilities (what is owed), and owner equity. Respondents were asked to provide an estimate of their non-real estate and real estate assets and liabilities. We wish to offer a word of caution about the reliability of the asset values reported. The value of total assets is the average amount reported by each respondent; we employed no mechanism for checking the accuracy of asset values among respondents. Hence the real estate and farm machinery values that made up a large portion of each respondent's balance sheet are subject to possible biases on the part of the respondents.

The average value of total farm assets reported by respondents as of January 1, 1989 was $310,972, ranging from $282,794 in the lakes states to $373,657 in the plains (Table 3.2). Farm real estate accounted for about 69 percent of total farm assets regionwide. The median values were substantially smaller than the mean values (median total assets ranged from 68 percent to 71 percent of mean values) and may be more representative of the situation for a typical producer. About 3.5 percent of the respondents regionwide reported that their total assets exceeded $1 million in value; about 20 percent had assets totaling less than $100,000. Farm non-real estate assets averaged $96,267 in value for the region.

For all respondents, the average amount of total debt was $95,103, ranging from $114,119 in the plains to $87,529 in the lakes states (Table 3.3). Real estate debt averaged $65,920; non-real estate liabilities averaged $29,183 regionwide.

The respondents' owner equity (total assets minus total liabilities), which averaged $215,870 for the region, was highest in the plains ($259,539) and lowest in the lakes ($195,265) (Table 3.4).

Table 3.5 displays the percentage of all respondents carrying real estate and non-real estate debt in 1989. Note that about 28.1 percent of all respondents reported zero debt in 1989. This proportion ranged from 24.2 percent in the lakes to 30.5 percent in the corn belt. To represent the

TABLE 3.2 Balance Sheet Information: Assets by Type for Respondents, 1988

Asset	Corn Belt	Plains	Lakes	NC Region
	means			
Real estate assets	$207,112	$250,875	$195,769	$214,705
Non-real estate assets	87,991	122,782	87,025	96,267
Total assets	$295,103	$373,657	$282,794	$310,972
	medians			
Real estate assets	$130,000	$164,800	$145,005	$150,000
Non-real estate assets	50,000	75,000	50,000	50,000
Total assets	$200,000	$265,000	$200,000	$210,000
Frequencies – Real Estate Assets				
	percent			
< $100,000	35.4	29.7	33.4	33.5
$100,000 - $199,999	29.1	24.4	32.4	28.9
$200,000 - $299,999	16.2	15.7	17.2	16.3
$300,000 - $499,999	10.1	16.8	9.6	11.6
$500,000 - $1,000,000	7.7	11.6	6.5	8.3
> $1,000,000	1.6	1.8	0.9	1.5
Frequencies – Non-real Estate Assets				
< $100,000	70.1	57.9	67.4	66.3
$100,000 - $199,999	19.5	23.0	20.2	20.5
$200,000 - $299,999	5.3	10.0	7.1	6.9
$300,000 - $499,999	3.3	5.1	3.9	3.9
$500,000 - $1,000,000	1.5	3.5	1.4	2.0
> $1,000,000	0.4	0.5	0.1	0.9
Frequencies – Total Assets				
< $100,000	23.3	16.8	18.6	20.4
$100,000 - $199,999	25.6	18.5	29.5	24.9
$200,000 - $299,999	20.0	19.1	20.5	19.9
$300,000 - $499,999	15.6	21.8	17.5	17.7
$500,000 - $1,000,000	12.3	18.6	11.1	13.5
> $1,000,000	3.1	5.1	2.9	3.5

TABLE 3.3 Balance Sheet Information: Debts by Type for Respondents, 1988

Debts	Corn Belt	Plains	Lakes	NC Region
	----------means----------			
Real estate liabilities	$64,436	$ 77,189	$58,510	$65,920
Non-real estate liabilities	25,283	36,930	29,019	29,183
Total liabilities	$89,719	$114,119	$87,529	$95,103
	----------medians----------			
Real estate liabilities	$ 9,000	$20,000	$18,000	$14,000
Non-real estate liabilities	0	$ 5,000	$ 2,000	$ 2,000
Total liabilities	$24,700	$42,000	$40,000	$32,000
Frequencies – Real Estate Debt				
	----------percent----------			
No real estate debt	46.3	40.6	41.8	43.6
< $50,000	24.2	22.0	21.4	22.9
$50,000 - $99,999	11.7	12.3	16.7	13.3
$100,000 - $199,999	9.4	12.8	12.3	11.0
$200,000 - $299,999	4.0	5.3	4.5	4.4
$300,000 - $499,999	2.6	4.8	2.6	3.2
≥ $500,000	1.8	2.2	0.7	1.6
Frequencies – Non-real Estate Debt				
	----------percent----------			
Non-real estate debt	50.0	45.1	47.9	48.2
< $50,000	35.8	29.8	33.9	33.8
$50,000 - $99,999	7.2	12.0	10.3	9.2
$100,000 - $199,999	5.3	9.8	5.6	6.5
$200,000 - $299,999	0.9	1.5	1.6	1.2
$300,000 - $499,999	0.4	1.3	0.4	0.6
≥ $500,000	0.5	0.4	0.4	0.4
Frequencies – Total Debt				
No debt	32.0	24.2	29.8	29.5
< $50,000	30.4	27.7	24.5	28.1
$50,000 - $99,999	13.7	13.3	16.2	14.3
$100,000 - $199,999	11.8	15.6	16.3	14.0
$200,000 - $299,999	5.6	6.9	7.5	6.4
$300,000 - $499,999	3.7	8.3	3.9	4.9
≥ $500,000	2.8	4.0	1.9	2.8

TABLE 3.4 Balance Sheet Information: Owner Equity and Debt-Asset Ratio

Item	Corn Belt	Plains	Lakes	NC Region
Equity				
Mean	$205,384	$259,539	$195,265	$215,870
Median	$141,000	$180,000	$140,000	$148,000
Frequencies				
< $100,000	37.2	30.6	37.0	35.5
$100,000 - $199,999	26.8	24.1	29.3	26.8
$200,000 - $299,999	15.7	15.6	15.8	15.7
$300,000 - $499,999	10.8	16.2	10.3	12.0
$500,000 - $1,000,000	7.6	10.5	5.8	7.8
> $1,000,000	2.0	3.1	1.9	2.2
Debt-Asset Ratio				
Mean	30.1	32.2	32.7	31.3
Median	15.2	22.0	18.9	17.7
Frequencies				
No debts	32.1	24.3	29.8	29.5
0% - 40%	41.1	43.1	38.1	40.8
40% - 70%	16.5	20.0	19.3	18.1
70% - 100%	6.7	8.9	7.7	7.5
> 100%	3.6	3.8	5.1	4.1

TABLE 3.5 Percentages of Respondents with Debt

Item	Corn Belt	Plains	Lakes	NC Region
Percentage of All Respondents with:				
Real estate debt	59.6	61.7	64.5	61.8
Non-real estate debt	58.2	56.7	60.9	58.4
No real estate or non-real estate debt	30.5	29.2	24.2	28.1
Percentage of Respondents with Gross Farm Sales of ≥ $100,000 with:				
Real estate debt	79.6	80.9	78.9	79.8
Non-real estate debt	72.7	73.0	73.1	72.9
No real estate or non-real estate debt	11.4	14.4	12.9	13.0
Percentage of Respondents with Gross Farm Sales of < $100,000 with:				
Real estate debt	51.8	53.7	56.2	53.7
Non-real estate debt	52.4	50.1	53.5	51.8
No real estate or non-real estate debt	37.7	34.9	30.4	34.7

financial situation more accurately, however, the respondents should be divided into part-time and full-time farmers. The percentage of respondents with $100,000 or more of gross farm sales carrying zero debt in 1989 is considerably lower than the percentage of all respondents; only about 13 percent of the full-time farmers reported zero debt in 1989. This figure ranged from 11.4 percent in the corn belt to 14.4 percent in the plains.

Debt-Asset Ratios

Solvency measures describe the amount of money a farmer would have remaining after all assets are converted to cash and all debts are retired. Solvency ratios measure the relationship between claims on the business (liabilities) and either total assets or owner equity. The debt-asset ratio is one such solvency ratio, which is calculated as follows:

$$\frac{\text{Total liabilities}}{\text{Total assets}} \times 100.0$$

Before discussing the debt-asset ratios obtained in this survey, we suggest caution about the reliability of this ratio as an indicator of farmers' financial condition. First, the total debt component of the ratio does not take into account how the debt is structured, which can influence a farmer's ability to service and repay debt. Second, one should keep in mind the problems discussed earlier, about the difficulty of establishing a value for farm assets. Finally, the change in the amount of owner equity can be the result of a profit or a loss in the previous year and/or the result of an increase or a decrease in the asset values. Without an income statement and the knowledge of asset values on the previous balance sheet, it is difficult to identify the reasons for the change in owner equity for an individual operation.

The average debt-asset ratio for the region was 31.3 percent as of January 1, 1989 (Table 3.4). The ratio was highest in the lakes (32.7 percent) and lowest in the corn belt (30.1 percent).

Harrington (1985) and other authors of U.S. Department of Agriculture (USDA) publications have employed debt-asset ratios to describe the amount of financial stress facing farmers, as follows:

Debt-Asset Ratio	Farmer's Status
Under 40%	No apparent financial problems
40% - 70%	Serious financial problems
71% - 100%	Extreme financial problems
Over 100%	Technically insolvent

About 30 percent of respondents in the North Central region had debt-asset ratios exceeding 40 percent in 1989; 11.6 percent had ratios exceeding 70 percent (Figure 3.3). The guidelines in the USDA classification scheme suggest that about 11.6 percent of all respondents in the region face extreme financial problems or technical insolvency; this interpretation, however, tends to overstate the problem. For example, some skilled managers who carry a relatively small proportion of their debt in the form of land debt may be in satisfactory financial condition even though they have a debt-asset ratio exceeding 70 percent. Later we will report additional analyses involving cross-tabulation and subsets of respondents to assess more fully the meaning of the debt-asset ratio figures.

The percentage of highly leveraged producers is highest in the plains (32.7 percent had debt-asset ratios of 40 percent or more) and

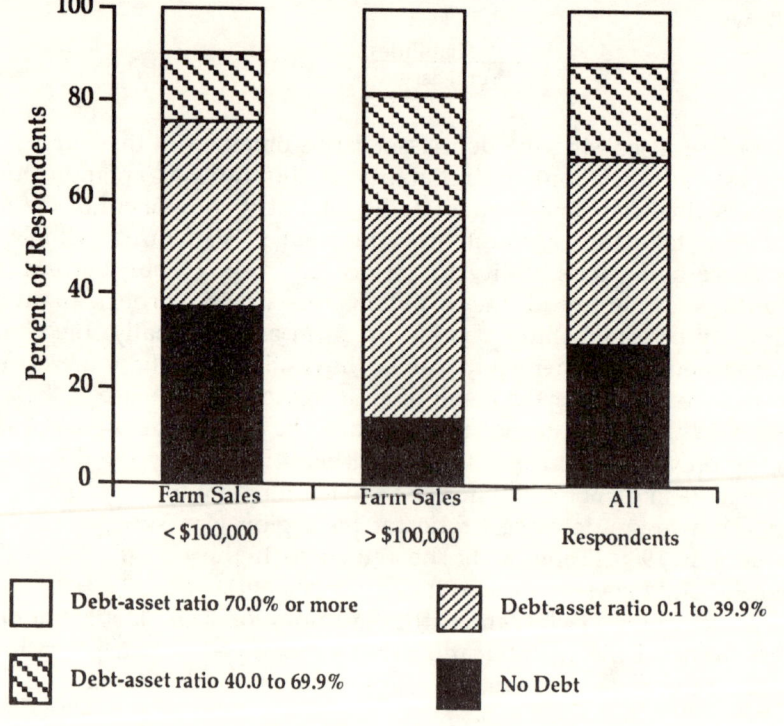

FIGURE 3.3 Degree of financial leverage for respondents with gross farm sales less than $100,000, sales of $100,000 or more, and all respondents, North Central Region, 1988

lowest in the corn belt, where only 26.8 percent had debt-asset ratios exceeding 40 percent (Table 3.4).

The percentage of highly leveraged operators is greater among farmers with gross incomes of $100,000 or more per year than among others (Table 3.6). Regionwide about 41 percent of these full-time farmers had debt-asset ratios of 40 percent or more, compared to 24.6 percent of producers with gross sales of less than $100,000. This pattern also held among the subregions. In the corn belt, 39.5 percent of producers with gross farm sales of $100,000 or more had debt-asset

TABLE 3.6 Distribution of Farms According to Debt-Asset Ratio for All Farmers in Surveys and for Farmers in Surveys with Gross Farm Sales Less Than and More Than $100,000 per Year

Debt-Asset Ratio Category	Percent of Respondents in Category Based on Figure for		
	All Farms	Farms with Gross Farm Sales	
		< $100,000	> $100,000
	------percent------		
Debt-Asset Ratio (NC Region)			
No debt	29.5	36.1	13.9
< 40.0%	40.8	39.1	45.0
40.0 - 69.9%	18.1	15.5	23.9
70.0 - 99.9%	7.5	5.7	11.6
≥ 100.0%	4.1	3.4	5.6
Debt-Asset Ratio (Corn Belt)			
No debt	32.1	37.5	15.9
< 40.0%	41.1	40.0	44.6
40.0 - 69.9%	16.5	13.9	23.7
70.0 - 99.9%	6.7	5.4	10.3
≥ 100.0%	3.6	3.0	5.5
Debt-Asset Ratio (Plains)			
No debt	24.3	31.2	13.9
< 40.0%	43.1	40.5	47.0
40.0 - 69.9%	20.0	17.8	23.4
70.0 - 99.9%	8.9	6.9	11.7
≥ 100.0%	3.8	3.6	4.0
Debt-Asset Ratio (Lakes)			
No debt	29.8	37.5	10.8
< 40.0%	38.1	36.3	43.3
40.0 - 69.9%	19.3	16.8	24.7
70.0 - 99.9%	7.7	5.4	13.4
≥ 100.0%	5.1	4.1	7.8

ratios exceeding 40 percent, compared to 22.3 percent of those whose gross sales were less than $100,000. The comparable values for the plains were 39.1 percent and 28.3 percent; for the lakes the values were 45.9 percent and 26.3 percent.

Delinquency Rates

Another measure of financial condition is the rate of delinquency of loan payments. Respondents having real estate loans and providing usable responses (39.4 percent of all respondents) were asked whether their principal and interest payments were current. For all respondents, 9.5 percent said "no" (Table 3.7). The same question was asked about non-real estate loans; about 14 percent of the respondents having non-real estate loans stated that their principal and interest payments were not current (Table 3.7). For both types of loans, the delinquency rates were highest in the corn belt.

Anticipated Financing Problems

A third indicator of farmers' financial condition is the percentage

TABLE 3.7 Delinquency Rates and Anticipated Financing Problems

Item	Corn Belt	Plains	Lakes	NC Region
Percentage of operators who have farm real estate mortgages	35.9	44.1	41.5	39.4
Percentage who are not current on mortgage payments	10.6	8.0	9.0	9.5
Percentage who have non-real estate loans	31.2	38.2	36.7	34.3
Percentage who are not current on non-real estate loans	14.7	13.0	12.4	13.7
Percentage who anticipate applying for new or expanded loans	19.6	22.5	18.9	20.1
Percentage who anticipate problems securing financing	5.9	7.1	6.9	6.5

of loan applications turned down by the lender. Respondents were asked whether they anticipated applying for a new or expanded farm loan in 1989; about 20 percent stated that they did so (Table 3.7). Those producers then were asked whether they expected any problems in securing financing. About 6.5 percent of the respondents thought they might experience such difficulty.

Amount of Debt Owed by Respondents in Different Debt-Asset and Gross Farm Sales Categories

Debt is concentrated in the hands of respondents in the higher debt-asset ratio categories. As noted in the figures for 1989, about 37 percent of the debt was owed by respondents with debt-asset ratios of 70 percent or higher; about 18 percent of the debt was owed by respondents who were technically insolvent. The 5.8 percent of the technically insolvent respondents and some respondents in the 70.0-to-99.9 percent debt-asset ratio category presumably were vulnerable to any financial adversities they might encounter.

Debt-Asset Ratio Category	Percentage of Respondents	Percentage of Debt
Under 40.0%	57.9	30.7
40.0 - 69.9%	25.7	32.4
70.0 - 99.9%	10.6	18.9
100.0% and over	5.8	18.0
Total	100.0	100.0

Debt also is concentrated in the hands of farmers with larger operations. In 1989, respondents with gross farm sales of $100,000 or more per year owed 68.8 percent of the debt, even though they made up only 30.8 percent of all respondents who answered the question.

Summary and Implications

The results reported here illustrate clearly the need to segregate the data used to report financial characteristics. The data should be segregated by geographic area and farm size because the degree of reliance on income from farming versus off-farm employment differs across geographic area and size category. In addition, financial characteristics vary across farm sizes.

The results illustrate the diversity across the North Central region in regard to percentages of the farm population that would be classified as part-time and full-time farmers. The percentages of full-time farmers for the corn belt and the lakes subregions are similar, whereas the percentage of full-time farmers is substantially higher in the plains.

The percentage of total family income that comes from farming varies only slightly across the subregions for operations with gross farm sales of $100,000 or more (7 percentage points), but the variation is relatively large (21 percentage points) for operations with gross farm sales of less than $100,000. The percentage of total family income from farming for operations with gross farm sales of less than $100,000 was only 42 percent in the corn belt and 49 percent in the lakes subregion, but 63 percent in the plains.

The proportion of all respondents carrying no debt was about 30 percent in 1989. For respondents with gross farm sales of $100,000 and over, however, that proportion was only 14 percent. Among respondents with less than $100,000 gross farm sales, the percentage with no debt was 36 percent.

The debt-asset ratio for the region as of January 1, 1989 was 31 percent, with little variation across the subregions. A more meaningful figure, however, is the percentage of respondents with debt-asset ratios in various debt-asset categories. The proportion of all respondents with debt-asset ratios in the 70 percent-and-over category varied little across the subregions (2.5 percentage points). The variation in the percentage, however, was larger across size groups. The proportion was more than 8 percentage points higher for full-time farmers than for part-time farmers in the North Central region. In the lakes subregion, the proportion of part-time farmers with debt-asset ratios of 70 percent and over was 9.5 percent, compared to 21.2 percent for full-time farmers.

These data show that the use of average figures for the country or for the North Central region does not represent farmers' financial condition adequately. Their financial characteristics vary across areas of the region and across farm size. It would follow logically that the responses to financial stress in the farm sector would differ across farming areas and size categories.

References

Dobson, W. D., F. L. Barnard, and B. Graves. 1985. *Results of Indiana Farm Finance Survey*. West Lafayette, IN: Purdue University, Department of Agricultural Economics.

Doye, D. G., R. W. Jolly, and D. Choat. 1987. "Agricultural Restructuring Requirements by Farm Credit System District." *Staff Report 87-SR34.* Ames: Iowa State University, Center for Agricultural and Rural Development.

Harrington, David H. 1985. "A Summary Report on the Financial Condition of Family-Size Commercial Farms." *Agricultural Information Bulletin 492.* Washington, DC: U.S. Department of Agriculture.

Johnson, J. D., M. J. Morehart, and K. Erickson. 1987. "Financial Conditions of the Farm Sector and Farm Operators." *Agricultural Finance Review* 47: 1-18.

Leholm, A. G., F. L. Leistritz, B. L. Ekstrom, and H. G. Vreugdenhil. 1985. "Selected Financial and Other Socioeconomic Characteristics of North Dakota Farm and Ranch Operators." *Agricultural Economics Report 199.* Fargo: North Dakota State University.

Leistritz, F. L. and B. L. Ekstrom. 1988. "The Financial Characteristics of Production Units and Producers Experiencing Financial Stress." Pp. 73-95 in *The Farm Financial Crisis: Socioeconomic Dimensions and Implications for Producers and Rural Areas,* edited by S. H. Murdock and F. L. Leistritz. Boulder: Westview.

Leistritz, F. L. and S. H. Murdock. 1988. "Financial Characteristics of Farms and of Farm Financial Markets and Policies in the United States." Pp. 13-28 in *The Farm Financial Crisis: Socioeconomic Dimensions and Implications for Producers and Rural Areas,* edited by S. H. Murdock and F. L. Leistritz. Boulder: Westview.

4

The Process of Adaptation and the Consequences to the Farm System

Bruce Johnson and Raymond D. Vlasin

Those engaged in production agriculture are acutely aware of the ever-changing world in which they operate. Change and the adaptation to change are the rule, not the exception. In fact, the ability to anticipate and assess dynamic conditions and to activate an effective response is fundamental to success. Thus the farm operator and his or her family tend to be adept at adjusting, including coping with difficult situations.

The farm financial crisis of the mid-1980s, however, was a period of unparalleled stress for farmers. Particularly in the North Central region, the reductions in income levels, the deep declines in asset values, and the debt-servicing problems were so pervasive that the sustainability of most farm units was in doubt. Except for older members of the community, who had vivid memories of the Depression years, few were ready for the "economic storm." In this environment more drastic, sometimes traumatic, responses were necessary. In addition to making family adjustments to cope with the financial crisis, operators and their families had to make considerable changes in their farming operations to ensure sustainability or orderly termination. As a part of this risk-reducing strategy, farming practices and resource use were altered in a variety of ways; these are the focus of this chapter.

The risk-reduction measures taken by farm families cover a full continuum from relatively minor steps, such as paying closer attention to marketing and maintaining more complete financial records, to seeking training for a new vocation and even terminating farming. Some measures represented the rather natural response of improving aspects of management which already were being employed; others were far more deliberated and required a significant change in the farming operation. It would be logical to expect that most farm families moved along this adjustment continuum over time, making the more obvious adjustments before taking more drastic measures. Thus one would expect this risk-reduction behavior to involve some combination of increasingly strong measures, much as medical treatment might be prescribed for a physical ailment.

The pattern of adjustments reflects not only the degree of impact of the financial crisis on a particular farming unit but also the nature and diversity of the environment in which that farm exists. Consequently one would expect differences in adjustment response, depending on the area of the country, the type and size of farm, various characteristics of the operator, off-farm opportunities, and other factors. Are the adjustments that were made rational and consistent? More specifically, to borrow from a farm management concept, are the adjustments complementary, supplementary, or competitive? The nature of the North Central Regional Farm Study precludes a definitive response to this question, largely because the farm-level adjustment process simply could not be studied in depth. Also, appropriate risk-reducing adjustments made by one farm may be inappropriate or counterproductive for another. For example, renting more acres of land may have been an effective means of reducing risk for one producer, while during the same period another producer was reducing risk by renting fewer acres. Consequently both responses appear on the inventory of adjustments made by farmers in the North Central region.

Finally, "risk reduction" can and does mean different things to different people. As a result, some of the strategies and tactics reported here may well be "risks" in themselves. Starting a new non-farm business is certainly a high-risk venture. Even diversifying a farming operation by adding a new crop or livestock enterprise can increase the risk, at least in the short run. Yet the operators considered such measures risk-reducing if they perceived them as contributing to the long-run resiliency and sustainability of the total farming operation and/or the economic well-being of the farm family.

Individual Adjustments

Farmers across the North Central region were asked what changes they had made in farming practices in the past five years to reduce risk. A number of actions were found to be widespread. For example, the farm financial crisis prompted a near-universal application of more careful marketing practices (Table 4.1). About four of every five farm operators in the region reportedly were making an effort to market their production more effectively. Regardless of the characteristics of the farming operation or the demographic traits of the operators, the great majority of respondents stated that they had taken action to improve marketing. They did not furnish details, however. For example, only a small percentage of respondents said they had begun to use futures markets to hedge commodity prices. Obviously, a number of other elements also were included in the efforts to improve marketing. Nevertheless, the producers' heightened awareness of the importance of marketing strategy is itself a positive manifestation of the economic stresses that agriculture faced during the 1980s. Insofar as permanent adjustments were made, a more resilient agricultural production system, responsive to market dynamics, is functioning in the 1990s.

Other farm-level adjustments also were reported by a majority of operators across the North Central region. More than seven of every ten operators said they had deliberately postponed major farm purchases during the past five years to reduce risk. When major purchases involved the use of credit, servicing obligations are expanded and financial risk is increased. Even when such purchases can be made from the operator's liquid assets, cash flow problems still are possible.

Deferral of large purchases is a common short-run tactic used throughout the business world. Such a tactic, however, is employed to "buy time." In production agriculture, as in any business, some purchases can be deferred only so long. Machinery and equipment replacement is a classic example. A considerable literature documents the aggregate reduction in value of machinery inventory in the farming sector during the last half of the 1980s. Farmers continued to operate with fully depreciated, often worn-out equipment during those years of financial uncertainty and economic stress. In time, replacement is necessary, and we have abundant evidence that replacement has been relatively heavy in the last few years of the 1980s and the early 1990s.

Servicing of agricultural debt became a difficult, if not insurmountable, task for agricultural producers in the early years of the decade—the result of rapid debt expansion during the 1970s followed by reduced income levels, diminishing asset values, and record-level interest rates. Agricultural producers quickly changed strategies from heavily debt-leveraged expansion to debt reduction and minimization of financial exposure. The lessons of the 1920s and the 1930s were being learned anew as shown by the results of the North Central Regional

TABLE 4.1 Farm Operators' Report of Risk-Reduction Tactics for 1984-1988, North Central Region and Subregions[1]

	Lakes States	Corn Belt States	Plains States	North Central States
	------------------------percent--------------------			
Paid closer attention to marketing	76	77	86	79
Postponed major farm purchase	71	71	74	72
Reduced long-term debt	61	65	68	65
Kept more complete financial records	63	61	66	63
Reduced short-term debt	60	62	65	62
Shared labor or machinery with neighbors	47	46	47	46
Reduced expenditures for hired help	42	43	43	43
Bought crop insurance	37	34	51	39
Diversified farm by raising livestock	33	37	41	37
Sought off-farm employment	33	36	26	33
Reduced machinery inventory	24	27	24	26
Rented more acres	22	23	27	24
Rented fewer acres	22	21	21	21
Diversified farm by adding new crops	22	15	28	20
Bought additional land	16	17	22	18
Used futures markets to hedge prices	14	19	16	17
Started a new business (not farming)	11	10	10	10
Changed from cash rent to crop share	6	12	13	10
Retired from farming	10	11	8	10
Sold some land	9	9	8	9
Sought training for a new vocation	9	8	6	8
Quit farming	7	8	5	7
Transferred land back to lender	3	5	5	4

[1]Lakes states include Michigan, Wisconsin, and Minnesota. Corn belt states include Ohio, Indiana, Illinois, Iowa, and Missouri. Plains states include North Dakota, South Dakota, Nebraska, and Kansas.

Farm Survey. More than 60 percent of the respondents reported efforts to reduce levels of both short-term and long-term debt. If one also considers those farmers who reduced only long-term debt and those who reduced only short-term debt, the combined proportion of farmers who reduced their debt loads would be considerably higher than 65 percent.

Regardless of the size of the agricultural operation or the farm family's dependence on the earnings of the operation, most operators made a deliberate adjustment between 1985 and 1989 to reduce their debt load. Liquidating assets, foregoing certain purchases, and accelerating debt repayment schedules were common tactics. In some instances, operators negotiated with lenders to write down the debt to more manageable levels. Throughout the United States, the agricultural production sector was rapidly purging debt; by the end of the 1980s, aggregate debt was just two-thirds of the level recorded five years earlier. This action contributed greatly to the economic recovery of the industry and to the reduction of financial risk borne by farm families.

The other risk-reducing measure reported by the majority of farm operators was the effort to maintain more complete financial records. Obviously this is a relative measure of change; "more complete" could range from an increase in detail beyond the minimum required for filing tax returns to the use of a highly sophisticated record system that was expanded to incorporate more refined monitoring and analysis. Moreover, in some instances such an adjustment was not the operator's active decision but was mandated by lenders or other third parties. Nevertheless, the improvement of farm financial record keeping is a step toward more effective management in a high-risk environment. It is hoped that this improvement was not merely a short-run adjustment to the farm crisis, which will fade as economic conditions return to normal.

Another tactic reportedly used to reduce risk was resurrected from earlier eras—sharing labor and machinery with neighbors. As new technology has evolved and as management has grown more sophisticated, farm operators have tended to move toward greater independence within the farming community. By substituting capital for labor, the individual operator has been able to accomplish much more with only the human capital of the immediate family. Moreover, the sheer size of today's farming operations often precludes the sharing of important pieces of machinery and equipment because the timing of operations is particularly critical for larger farms.

The trend to independence in use of labor and machinery was not without cost, however, and the farm crisis allowed that cost to manifest itself in the form of burdensome debt on heavy capital investment. Consequently some operators began to share both machinery and labor to reduce their financial risk and sustain their farming operations. Nearly half reported using this tactic between 1985 and 1989.

The incidence of sharing varied substantially by type of operator and farming experience. The practice was used much more frequently by younger operators and/or operators who had been farming for only a short time (see Table 4.2). In many of these instances, the sharing probably occurred among family members operating farms nearby. It is likely that sharing of labor and equipment also involves some compensation for services provided. Such a tactic can be effective when used to delay costly purchases of capital equipment, as discussed above.

As the financial management of farmers increases in sophistication, the high fixed-cost investment in certain machinery increases. Some operators begin to realize that they simply do not have enough volume to justify such an investment. Thus, by sharing certain pieces of machinery with neighbors they can reach sufficient volume to justify the investment. The cooperating operators can be competitive in terms of economies of size. In short, such a tactic can be a "win-win" situation for all parties concerned. Questions remain, however: whether the social dynamic of the community would remain sufficiently intact to allow a more interdependent farming system and whether the existing structure of production agriculture still would be conducive to such interaction, except in crisis conditions.

Complementing the shift to greater sharing of labor and machinery with neighbors is a second pervasive adjustment, namely the reduction in expenditures for hired labor. More than two-fifths of the farmers reported such reduction. Both the reduction of expenditures for hired labor and the sharing of labor and machinery seemed to be uniform across the North Central region. Reductions were most frequent among the farmers in the 35-to-44 age group, possibly because more young family members were available to work in production. As noted in Table 4.1, numerous other steps of risk reduction and adaptation were taken by farm operators in the North Central region, although with less frequency and/or variability across subregions. For example, more than half of the operators in the plains states reported buying crop insurance to reduce risk in their operation whereas the incidence was 37 percent in the lakes states and 34 percent in the corn belt. This pattern appears logical because climatic conditions, and hence crop yields, can vary much more widely in the plains than elsewhere. Crop insurance makes greater economic sense for the North Dakota wheat producer than for the Illinois corn and soybean farmer.

Another change that varied considerably across the subregions was the frequency of seeking off-farm employment. About one-fourth of the

TABLE 4.2 Farm Operators' Report of Risk-Reduction Tactics for 1984-1988, by Age of Farm Operator, North Central Region

	Operator's Age (Years)			
	Under 35	35-44	45-64	65 or Over
	------------------percent------------------			
Paid closer attention to marketing	85	82	80	68
Postponed major farm purchase	76	79	73	60
Reduced long-term debt	64	65	68	55
Keep more complete financial records	76	71	60	53
Reduced short-term debt	64	67	64	49
Shared labor or machinery with neighbors	59	52	44	37
Reduced expenditures for hired help	41	47	43	37
Bought crop insurance	46	43	39	31
Diversified farm by raising livestock	44	41	36	31
Sought off-farm employment	43	43	33	15
Reduced machinery inventory	17	24	27	29
Rented more acres	45	32	20	8
Rented fewer acres	17	22	21	22
Diversified farm by adding new crops	26	25	19	13
Bought additional land	26	23	17	10
Used futures markets to hedge prices	24	21	16	10
Started a new business (not farming)	14	15	10	4
Changed from cash rent to crop share	13	14	9	7
Retired from farming	--	1	7	30
Sold some land	3	9	9	13
Sought training for a new vocation	12	11	6	3
Quit farming	2	4	6	15
Transferred land back to lender	5	6	5	3

respondents in the plains states had taken this step to reduce risk and adapt to the crisis. In contrast, the incidence of seeking off-farm employment was one-third or more in the other parts of the region. The differences observed here may reflect the lesser availability of off-farm employment and the greater commuting distances to urban centers in the more sparsely populated plains states.

A minority of farm operators across the region reported a number of changes in the operation itself. Adjustments in the land base (both rental and ownership), reduction of machinery inventory, and diversification of the farm by adding new crop or livestock enterprises were fairly frequent. Obviously the stressful financial conditions of the mid-1980s required rather significant changes. As noted previously, these changes often differed dramatically from one farm to the next. Across the region, for example, about one-fourth of the farm operators reported renting more land. In this way they acquired control of land without the major financial outlay associated with purchase. More than one-fifth of the operators, however, reported renting fewer acres to reduce risk and adapt to the current financial conditions. These farmers probably made this decision in order to reduce the rental acres that represented marginal, if any, economic benefit to the overall farming operation, and thus to concentrate more effectively on managing the remaining land. In some instances the reduction of rental acreage may have been combined with increased off-farm employment for the operator, the spouse, or other family members. In these cases, the decisions, though possibly opposed to one another, probably were economically rational.

For a small percentage of farmers, the farm financial crisis dictated more drastic measures whereby they reduced their economic dependence on farming to varying degrees. One farmer in ten attempted to start a new (non-farm) business; a somewhat smaller percentage stated that they had sought training for a new vocation. These farmers were taking steps to adapt to other careers. This survey could not determine whether those steps were taken in anticipation of a partial or a complete severance from farming. Clearly, however, this action represented a portion of the region's farmers who were willing to look beyond the farm gate and to recognize that their skills and experience could transfer to non-farm alternatives. They pursued a major change in career, like millions of Americans in other sectors of the economy during any recession. Although a change of this magnitude creates risk itself, it also represents an attempt to move to more sustainable and more stable economic conditions in the long run.

How does the operator's age affect the risk-reduction strategies? Initially it was proposed that younger farm operators more frequently than their older counterparts, would employ some of the more drastic measures of adaptation because of greater necessity as well as greater willingness to make significant change. Indeed, major differences in risk-reduction strategies were observed across the age groups (Table 4.2). The most frequently reported adjustments—attention to marketing, postponing major farm purchases, keeping more complete records, and reducing debt—were made commonly by all except the oldest age group, persons 65 or older. In contrast, the youngest farmers—those under age 35—shared labor or machinery with neighbors and sought off-farm employment more frequently than their older neighbors. Often with greater constraints on resources, they more frequently attempted to diversify their operations and expand their acreage by renting more land. They also reported the highest incidence using futures markets to hedge prices. In addition, the younger operators most frequently reported moving to change vocations. Indeed, the most significant changes apparently were made by the younger farmers. Even though farming was terminated most frequently in the oldest group, it represented a natural transition into retirement and usually was voluntary rather than a forced exit due to the farm financial crisis.

It would be logical to assume that the appropriateness of risk-reduction measures would vary with farm size. Using annual gross farm sales as an economic measure of farm size, we observed variations in incidence of use in The North Central Study (Table 4.3). More careful attention to marketing, reduction of long-term debt, and renting more acres became more frequent with the size of the operation. The incidence of seeking off-farm employment, however, was related inversely to farm size. This finding is quite logical in light of the availability of operator labor for such enterprises.

In short, it appears that operators of the larger farms tended to intensify and even expand their farming, while those on the smaller farms were more likely to reduce or even eventually terminate the operation. These differences are due to the relative availability of various resources and to the level of critical mass in terms of financial resources, land, and equipment necessary to operate an economically viable farm.

It would be logical to expect that the risk-reduction measures would tend to increase in frequency and magnitude as the operator's perceived stress level rises. The survey asked questions about the operator's stress, which allowed cross-tabulation of operator's stress level by tactics (Table 4.4). The most common risk-reduction tactics decreased measurably as farm operators saw their stress level remain the same or decline. Similarly, the more drastic measures were less frequent among operators who reported reduction in their stress levels. Unfortunately,

TABLE 4.3 Farm Operators' Report of Risk-Reduction Tactics for 1984-1988, by Economic Size of Farming Unit, North Central Region

	Annual Gross Farm Sales			
	Under $40,000	$40,000-99,999	$100,000-249,999	$250,000 and Over
	----------percent----------			
Paid closer attention to marketing	73	83	84	85
Postponed major farm purchase	69	77	73	74
Reduced long-term debt	60	66	70	71
Kept more complete financial records	57	67	69	67
Reduced short-term debt	57	65	68	64
Shared labor or machinery with neighbors	47	50	45	40
Reduced expenditures for hired help	41	47	42	43
Bought crop insurance	27	48	48	46
Diversified farm by raising livestock	36	40	37	36
Sought off-farm employment	46	31	19	12
Reduced machinery inventory	29	24	21	21
Rented more acres	12	25	37	47
Rented fewer acres	25	22	17	12
Diversified farm by adding new crops	18	21	24	25
Bought additional land	11	18	26	38
Used futures markets to hedge prices	12	17	20	37
Started a new business (not farming)	13	9	9	12
Changed from cash rent to crop share	12	9	10	11
Retired from farming	15	6	4	2
Sold some land	10	7	9	13
Sought training for a new vocation	10	7	7	4
Quit farming	11	4	2	2
Transferred land back to lender	4	4	6	9

however, this observed relationship begs the question of causality: did greater and increasing stress lead to more actions to reduce risk in the farming operation, or did the risk-reduction tactics employed actually lead to reduced stress by the end of the 1984-1988 period, and thus were reported as reducing stress? We suspect that some combination of these two phenomena was at work.

The most noteworthy feature of Table 4.4 is the difference in response between farm operators and their spouses. Many of the actions taken most frequently appeared to be more stress-relieving for the spouses than for the operators. One possible explanation is that the spouse often is more deeply involved with bookkeeping for the farm than the operator, who must "wear many hats" in the day-to-day conduct of the operation. As a result, the spouses often could see the initial financial deterioration more clearly, and were more anxious to take measurable steps toward resolution. The mere taking of such steps apparently was therapeutic and was an indication that all persons involved were realistic in assessing the financial threat.

Finally, we observed patterns of risk-reducing tactics by net family income level in 1988 (Table 4.5). Greater attention to marketing was high in all income groups, but particularly among those who still reported negative income for 1988. This group also reported the highest incidence of postponing major farm purchases, keeping more complete financial records, sharing labor and machinery with neighbors, and seeking off-farm employment.

The group of farmers that reported negative family income in 1988 also reported the highest incidence of reducing expenditures for hired help, purchasing crop insurance, seeking off-farm employment, reducing machinery inventory, and renting fewer acres. Also, they most often attempted to start a new business, changed from cash rent to crop share, sought training for a new vocation, and transferred land back to lenders. This group was highly active in taking measures to adjust to the financial crisis, but apparently to little avail, because they reported negtive family income for 1988. Aggregate farm family income in that year was at favorable levels, but these families, about 5 percent of the sur-vey respondents, were sinking into even deeper financial difficulties. Similarly, those whose 1988 income level was below $20,000 were relatively active in using adjustment tactics. Yet although their income was modest, it still may reflect considerable progress toward economic recovery.

At the other end of the family income continuum, nearly one-third of the respondents reported family income of $30,000 or more for 1988. The risk-reduction tactics employed by these operators followed the overall pattern, but often they were used somewhat less. Apparently

TABLE 4.4 Farm Operators' Report of Risk-Reduction Tactics for 1984-1988, by Reported Changes in Stress of Farm Operators' Spouses, North Central Region

	Changes in Operator's Stress Level				Changes in Spouse's Stress Level			
	Increased Greatly	Increased Somewhat	Same	Declined Somewhat	Declined Greatly	Increased Greatly	Increased Somewhat	Same	Declined Somewhat	Declined Greatly
	----percent----					----percent----			
Paid closer attention to marketing	87	84	67	69	61	85	82	76	76	75
Postponed major farm purchase	86	79	55	59	56	88	77	62	62	62
Reduced long-term debt	64	66	58	64	61	67	64	62	65	71
Kept more complete financial records	74	69	49	49	43	72	69	55	54	51
Reduced short-term debt	64	62	56	62	49	66	64	55	64	61
Shared labor or machinery with neighbors	53	49	38	39	48	55	50	40	42	49
Reduced expenditures for hired help	61	44	27	29	30	57	44	34	31	47
Bought crop insurance	48	39	28	28	27	46	39	34	34	39
Diversified farm by raising livestock	47	39	29	30	32	44	37	35	34	33
Sought off-farm employment	46	34	24	22	25	45	34	25	27	33
Reduced machinery inventory	37	26	17	23	26	32	26	20	22	19

Rented more acres	28	25	17	23	26	32	26	19	22	23
Rented fewer acres	30	21	16	19	26	25	22	18	20	17
Diversified farm by adding new crops	27	21	15	20	18	25	22	17	16	20
Bought additional land	18	19	17	19	9	14	20	20	21	15
Used futures markets to hedge prices	18	18	11	13	14	19	17	16	16	11
Started a new business (not farming)	17	9	7	8	14	14	11	6	9	5
Changed from cash rent to crop share	15	11	7	7	4	13	10	10	6	9
Retired from farming	8	8	12	13	20	7	8	11	12	12
Sold some land	13	8	7	11	9	13	9	6	13	9
Sought training for a new vocation	13	7	3	6	5	12	8	4	6	7
Quit farming	8	5	6	9	9	6	6	5	11	5
Transferred land back to lender	10	3	2	2	5	9	4	2	5	3

83

TABLE 4.5 Farm Operators' Report of Risk-Reduction Tactics for 1984-1988, by Net Family Income Level, North Central Region

			Net Family Income 1988			
	Loss	$1 - 9,999	$10,000 - 19,999	$20,000 - 29,999	$30,000 - 39,999	$40,000 or More
			percent			
Paid closer attention to marketing	86	74	78	79	84	80
Postponed major farm purchase	83	77	74	72	72	66
Reduced long-term debt	63	62	63	64	66	70
Kept more complete financial records	73	64	64	61	65	60
Reduced short-term debt	61	60	61	63	64	64
Shared labor or machinery with neighbors	54	43	48	48	49	43
Reduced expenditures for hired help	58	47	46	42	42	37
Bought crop insurance	43	42	41	41	38	33
Diversified farm by raising livestock	43	46	39	37	35	32
Sought off-farm employment	48	31	33	32	35	33
Reduced machinery inventory	39	31	29	21	25	20
Rented more acres	21	22	24	24	23	25
Rented fewer acres	31	24	22	19	22	19
Diversified farm by adding new crops	28	18	22	23	18	19
Bought additional land	12	15	15	16	21	26
Used futures markets to hedge prices	20	12	14	18	17	20
Started a new business (not farming)	19	8	10	11	9	12
Changed from cash rent to crop share	15	7	12	11	10	10
Retired from farming	9	14	11	10	6	5
Sold some land	8	9	9	9	8	10
Sought training for a new vocation	14	6	8	9	6	8
Quit farming	7	10	8	7	4	5
Transferred land back to lender	9	5	4	5	4	4

these farm families avoided at least some of the farm financial crisis and were enjoying a relatively favorable net family income in 1988.

Syntheses of Adjustments

In all, 23 risk-reducing actions were offered for consideration by the respondents. They can be classified as 1) those which increase income from farming or from non-farm sources or which improve efficiency, 2) those which reduce costs or postpone expenditures, 3) those which transfer resources out of farming, and 4) other actions that minimize exposure to risk.

Increasing income or improving efficiency was common throughout the North Central region. Nearly four out of every five operators reported giving closer attention to marketing; 63 percent stated that they kept more complete financial records, thus reinforcing their marketing efforts. A third of the operators reported seeking off-farm employment as a means of reducing their risk in farming; this figure is in addition to the significant number of operators already employed in off-farm jobs. About a fourth rented additional land; and 18 percent bought additional land. One in ten reported that they had started a new non-farm business.

Decreasing expenses and postponing expenditures was another common risk-reducing tactic. Among the most prevalent actions, 72 percent of the operators postponed major farm purchases, 65 percent reduced long-term debt, and 62 percent reduced short-term debt. In addition, 46 percent shared labor or machinery with neighbors, and 43 per-cent reduced expenditures for hired labor. About 26 percent reduced machinery inventory, and 21 percent reported reducing the number of acres rented.

Finally, as this analysis considers the total adjustment process, it becomes quite clear that individual farm families often took a number of steps simultaneously to adjust to the conditions facing them. Although a combination of adjustments frequently was required, it is reasonable to question whether the overall effort in fact was sustainable beyond the immediate crisis period. Specifically, many of the adjustments imply a sizable increase in the time and effort required by the farm family. Where will that additional time and effort come from? Might the resource requirements exceed a sustainable level? If they do so, could not the greater demands for time and attention place new or additional stress on the farm family?

Conclusions

Farm families across the North Central region faced difficult financial times during the 1980s, and responded with a full array of logical and sometimes creative adjustments. Generally these actions were related to farm and financial management issues designed to sustain the farming operation and to increase internal efficiency and economic viability. Less frequently, operators and their families felt it necessary to take more drastic measures that represented partial, if not eventually complete, termination of their farming operation.

Although some variation was observed from different perspectives, it is clear that farmers were making the necessary adjustments and were not denying the overall economic reality. Their actions, as shown by this survey, certainly would suggest a high level of resiliency and adaptability. Time will tell, however, whether these actions were sufficient to weather the economic storm of the 1980s and to prepare these farmers of the U.S. heartland for the future.

5

Plans for Changing the Farm Business and Needs for Training

Kent D. Olson and William E. Saupe[1]

In this chapter we address the changes that farm operators planned to make in their farm businesses in the five years following the North Central Regional Farm Survey, and we discuss the information and training that they perceived needing during that period. By comparing and contrasting their farm business plans with their needs for information and training, we provide additional insights useful to farm households, agricultural educational institutions, the agricultural input and marketing sectors, and others involved in the future of agriculture. We begin by reviewing the conceptual framework in which the analysis is conducted.

The Conceptual Framework

The content of this chapter fits well into the farm household/farm business/community concept presented in the lead chapter of this volume. The farm business contributes to the economic well-being of the farm household through the income it generates and through the potential contribution of any increase in the value of farm assets to the household's net worth. It also contributes to well-being if the household members prefer a rural place of residence, self-employment, and minimal travel time to the place of work. Later in this chapter we will show that some farmers may increase the contribution of the farm business to household income and well-being through their plans to add new crops or to raise livestock, or by renting or buying additional land. Others, however, plan to quit or retire from farming in the next five years, ending the farm business as a source of financial contribution to the household.

The farm household in turn provides labor to the farm business and to the non-farm labor market in the community in which it is located. In some cases, more family labor may be allocated to the farm business in the future if the farm business carries out its plans to reduce farm expenditures for hired help or to increase the on-farm processing of farm products before selling. On the other hand, more farm household labor will be supplied to the non-farm labor market if household members fulfill their plans to seek training for a new vocation, to seek off-farm employment, to start a new non-farm business, or to quit farming.

The community is linked structurally to farm households and to farm businesses. The local community may well be the labor market in which farm household members will seek employment and a market for the new crops or expanded livestock production that some farm businesses plan. The farm input businesses in the community will be affected if the farm's product mix is changed, if low-input production methods are adopted, if new production technologies are adopted, or if major farm purchases are postponed; all of these steps have been indicated in the plans by some farm businesses. The community infrastructure may be required to change to provide the vocational training that some farm operators want for non-farm employment, and the farm credit industry may be affected by farm businesses' plans to reduce debt. Thus, although this chapter pertains specifically to planned changes in the farm business and the needs for farm training perceived by the farmer, the effects also may be felt in the farm household and the community because of the extensive structural links and interrelationships.

The Audience

Knowledge of these future needs for training is important because agricultural production is a complex business. Successful farmers and ranchers have extensive needs for education and information. Agricultural producers must be skilled in production technology, business and financial management, marketing, and many other areas to be effective in the business world and in their households.

Agricultural producers develop these skills and abilities over time; they learn from their own experiences, from their relatives and neighbors, and from many other sources. A shorter learning period is less costly to the producer and to society in terms of income foregone, wasted resources, and human trauma. The public sector has shown that it has an interest in shortening the learning period: it provides structured learning through the Cooperative Extension Service, short courses and degree programs in colleges of agriculture, farm training programs in vocational and technical institutions, and high school agricultural courses.

The non-farm agricultural industry also has a stake. For example, agricultural lenders and farm business management associations provide instruction in farm business analysis and financial planning. The farm input industry provides technical assistance, training, and information relative to its products. Marketing institutions help farmers to understand the many new options available for marketing farm and ranch products.

Source of Data

Our analysis is based on the responses to two questions made by the 4,087 farm operators in the regional sample. These questions concerned changes that the respondents were planning to make in their farm businesses and their perceptions of their future needs for information and training.

In the first question (question 8B in the operator's questionnaire), the farm operators indicated the changes that they planned to make in their farm businesses in the next five years. This question read as follows: "Many farmers believe that risk in farming has increased during the past five years. Please indicate the changes you are planning to make in the next five years to reduce risk in your operation."

Respondents were asked to evaluate the likelihood of their making each of 23 possible changes. This question parallels the question discussed by Johnson and Vlasin in Chapter 4 of this volume, in which farmers indicated the changes they had made in their farm businesses in the past five years. In the tables and text that follow, we report the percentages of farmers and ranchers that said they were planning to make the indicated changes.

The second question (question 10 in the operator's questionnaire), inquired directly about the farmers' future needs for information and training. Farm operators were asked: "In order to continue farming in the next five years, I will need information or training on . . ." (followed by a list of nine farming practices).

Respondents were asked to rank each farming practice on a five-point scale on which 1=not needed, 2=low need, 3=moderate need, 4=high need, and 5=very high need. In the tables and text that follow, we have included only the operators who reported a high need or very high need for information or training.

Method of Analysis

Farming and ranching in the North Central region is highly diverse. That diversity affects the operators' farm business plans for the future and influences their perceptions of their needs for information and training. An understanding of that diversity can lead to a more appropriate response to the changing structure of farming made by rural communities, educators, the farm input and marketing sectors, and the farm households themselves.

In the analyses that follow, we first present the responses made by all 4,087 farm operators in our 12-state regional sample. Then we sort these responses by three aspects of diversity that may affect the operators' plans or perceptions. We also discuss the major differences in the responses.

Aspects of Diversity

Diversity within the North Central region. The 12 states included in the North Central region differ in climate, natural resources and man-made conditions such as urbanization. These differences may influence the changing structure of farming; accordingly the appropriate educational and other responses may vary in different parts of the region.

We report our results for the region as a whole and for the three subregions. The corn belt includes Ohio, Indiana, Illinois, Iowa, and Missouri, and is represented by a sample of 2,112 farmers. The plains states include North Dakota, South Dakota, Nebraska, and Kansas and are represented by 947 farm and ranch operators. The lakes states, Michigan, Wisconsin, and Minnesota, are represented by 1,028 farmers.

Diversity in operators' ages. Younger farmers, those under age 35, probably are also recent farm entrants. They make up 13.8 percent of all farms in our regional sample. This group has particular needs related to gaining control of farm assets and becoming established in farming. On the other hand, "pre-retirement" farmers, those age 60 and older, make up 32.0 percent of our regional sample. Their needs for information focus on the other end of the farming cycle—that is, when and how to phase out of farming.

Diversity in size of farm business. An inclusive definition of "farm" is used by the U.S. Department of Agriculture and the agricultural census. The current definition, in use since the 1974 census, includes any place from which $1,000 or more of agricultural products were produced and sold, or would have been sold, in a normal year. "Farms" and "farmers" thus range from relatively minor agricultural interests pursued by rural residents to businesses that are a major, or the sole, economic activity of the farm household.

We address the diversity of needs for information by contrasting farmers where gross annual sales of farm products are less than $10,000 (22.0 percent of our sample) with those whose sales are $100,000 or more (29.9 percent of our sample). This division also reflects their occupational diversity: on farms with relatively small gross sales, non-farm work is likely to be the major source of household income.

Table Format

In each of the following tables, the number and percentage of farmers in each portion of the sample are reported at the head of each column. The remainder of the column reports the percentage of that number of farms which has the particular attribute under examination. In Table 5.1, for example, 2,112 (or 51.6 percent) of all 4,087 farms are located in the corn belt. Farther down the column, we see that 11 percent of those 2,112 corn belt farms plan to diversify their farms by adding new crops, 26 percent by raising livestock, 13 percent by buying additional land, and so on.

Plans and Needs Reported by the Regional Sample

Plans for Changing the Farm Business

Four changes in farm businesses were planned for the near future by more than half of all farm operators in the survey: "pay closer attention to marketing" (67 percent), "keep more complete financial records" (57 percent), "reduce long-term debt" (54 percent), and "reduce short-term debt" (52 percent) (see the North Central Region column in Table 5.1). The fact that these four steps are planned by a high percentage of the respondents will surprise no one who is in close contact with farmers. These responses show that marketing, recordkeeping, and debt reduction will continue to be high priorities for many farms. They will be important to farmers, and thus to educators and the agricultural service industries as well.

TABLE 5.1 Percentages of All Sample Farmers in the North Central Region Planning to Make Selected Changes in the Near Future

	Corn Belt States	Plains States	Lakes States	NC Region
(Number of sample farmers)	(2,112)	(947)	(1,028)	(4,087)
Percentage of sample farmers	51.6	23.2	25.2	100.0
Diversification of Farm Enterprises				
By adding new crops	11	17	19	15
By raising livestock	26	28	23	26
Control of Land				
Buy additional land	13	18	10	13
Rent more acres	20	24	16	20
Rent fewer acres	14	14	15	14
Sell some land	4	4	7	5
Transfer land to lender	2	2	1	1
Operator's Involvement in Farming				
Seek training for new vocation	7	7	8	7
Seek off-farm employment	24	20	24	23
Start a new business (not farming)	7	7	9	8
Retire from farming	13	12	14	13
Quit farming	10	8	10	9
Reduction of Risk Exposure				
Reduce long-term debt	54	55	53	54
Reduce short-term debt	51	54	51	52
Use the futures market to hedge prices	18	19	16	17
Change from cash rent to crop share	11	12	6	10
Buy crop insurance	30	37	35	33
Reduction of Cash Expenditures				
Postpone major farm purchase(s)	43	44	46	44
Share labor or machinery with neighbors	34	35	34	34
Reduce expenditures for hired help	30	29	31	30
Reduce machinery inventory	18	19	17	18
Financial Management and Marketing				
Keep more complete financial records	56	58	59	57
Pay closer attention to marketing	66	71	65	67

Three other steps were planned by one-third or more of all farmers surveyed. Forty-four percent said they planned to postpone major farm purchases in order to reduce cash expenditures. This postponement could lead to slower growth in farm size, less growth in farm mechanization, and delays in replacing the current machinery. Perhaps to the detriment of a farm's future, some postponements may delay the adoption of new production and information technologies. This situation may damage competitiveness and the ability to adapt and respond to economic challenges.

"Sharing labor or machinery with neighbors" was planned by 34 percent of the farmers from the 12 states. This willingness to share labor can be coupled with the plans of 30 percent to "reduce expenditures for hired help." Together these two items show that farmers may expect to provide more of their farm labor themselves, to give up some of their independence, and to work more closely with their neighbors in order to reduce both their expenditures and their exposure to risk.

Thirty-three percent stated that they planned to buy crop insurance in the next five years as one way to reduce risk. We recall that the farmers were indicating the *changes* they planned to make in the next five years. That is, buying crop insurance would be a change in business practices for 33 percent of the operators, but buying crop insurance may be standard practice for another group. Thus the actual percentage of farmers planning to buy crop insurance can be expected to exceed 33 percent.

Diversification in the sources of farm family income is reflected in three responses. More than one-fourth of the farmers said they planned to diversify by raising livestock, an apparent reversal of past trends toward dropping livestock enterprises. Almost one-fourth said they planned to seek off-farm employment, a continuation of present trends. A smaller proportion, about 15 percent, said they planned to diversify by adding new crops.

Some responses were made by relatively small percentages of farmers. Because the sample represents 868,800 farm households in the region, however, a "small" percentage still reflects a large group of potential clients for extension and other programs. For example, the 13 percent who planned to retire from farming, the 9 percent who planned to quit farming, the 5 percent who planned to sell some land, and the 1 percent who planned to transfer some land to a lender total perhaps 100,000 farmers concerned with transferring or disposing of farm assets (if we assume some overlap among the respondents). The 8 percent who planned to start a new business and the 7 percent who planned to seek training for a new vocation represent about 25,000 farmers *each year* who could benefit from dislocated farmer programs, vocational-technical school counseling and training, and the array of services offered by state and federal job service programs.

Needs for Information and Training

When asked to indicate their needs for information or training, the largest group of farmers responded in the traditional interest areas of agricultural technology and farm marketing (see Table 5.2). One-third said they wanted information and training on "reducing production costs through low-input farming methods," 30 percent on "using new technologies as they become available," and 24 percent on "using new machines and chemical inputs to increase production." Although these items may seem contradictory, all three indicate the need for information on new farming methods and inputs.

Thirty-one percent of the farmers stated a need for information and training in marketing skills, another traditional educational need of farm producers.

When comparing changes in farming plans for the next five years with stated needs for education and training, we found that the farmers were not entirely consistent in their responses to the corresponding items in the two questions studied here. Two-thirds said they would pay closer attention to marketing, but slightly fewer than one-third said they needed training in marketing skills. While more than half planned to keep more complete financial records, fewer than one-fourth said they needed training. Hence, although many farmers recognize a need for training, many others plan to give more attention to farm records and marketing but do not mention the need for education or training. The latter group may need information about the availability of education and training opportunities, as well as encouragement to continue keeping more useful records and making better-informed marketing decisions.

Breakdown by Aspects of Diversity

Differences Within the North Central Region

Some noticeable differences exist among farmers in the corn belt, the plains states, and the lake states in their responses to the questions about plans for the future and needs for training. These differences are relevant because they have implications for the content and conduct of farmer education programs. In the corn belt, with its traditional corn-

TABLE 5.2 Percentages of All Sample Farmers in the North Central Region Reporting a Need for Information and Training

	Corn Belt States	Plains States	Lakes States	NC Region
(Number of sample farmers)	(2,112)	(947)	(1,028)	(4,087)
Percentage of sample farmers	51.6	23.2	25.2	100.0
Production Technology				
Reducing production costs through low-input farming methods	34	32	33	33
Using new technologies as they become available	29	31	32	30
Using appropriate conservation techniques	20	20	21	20
Business Management and Marketing				
Available government assistance	20	26	20	21
Marketing skills	33	33	28	31
Bookkeeping and financial systems	21	25	21	22
New Inputs and Enterprises				
Using new machines and chemical inputs to increase my production	23	24	25	24
Diversification of farm operation by adopting new crops and livestock	14	17	18	16
On-Farm Processing				
Processing farm products on farm before selling	9	11	9	10

soybean cropping system, fewer farmers planned to diversify by adding new crops. In the lakes states, where dairy farming and meat animal production are well established in capital-intensive facilities, more farmers planned to diversify by adding crops and fewer planned to change by raising livestock.

The plains states contain the smallest percentage of farmers who plan to seek off-farm employment, probably because fewer off-farm opportunities exist in the plains than in the corn belt and the lakes states. In another aspect of increasing employment, more of the plains farmers planned to buy or rent more land, whereas the lakes states contain the smallest percentage of farmers planning that course of action.

A larger percentage of farmers in the plains states planned to pay closer attention to marketing and to buy crop insurance. Those farmers also reported a need for information and training, both in available government assistance and in bookkeeping and financial systems (see Table 5.2).

One-third of the farmers in the corn belt and the plains states reported a need for information and training in marketing skills. In the lakes states, where dairy farmers have not been so dependent on markets for pricing their product, fewer farmers reported this need.

In keeping with the relatively small proportions of farmers who plan to diversify by changing crops, a smaller percentage of corn belt farmers reported a need for information and training on diversification of the farm.

Responses by Operator's Age

Table 5.3 displays the percentages of farmers planning to make selected changes in the next five years, as reported for two diverse age groups and the three subregions. We chose these two age groups to show the differences between the youngest farmers, who will be farming for the next 25 to 35 years, and the oldest, who probably will retire in the next ten years. The middle group is not forgotten; this breakdown serves only to highlight differences. The selected changes are grouped into six related clusters pertaining to increased diversity in farm enterprises, future control of land, changes in operator's involvement in farming, reduction in exposure to risk, reduction in cash expenditures, and changes in financial management and marketing.

In most instances, fewer of the older farmers in all subregions were planning to make the suggested changes. The major exceptions, as might be expected, are that more older farmers planned to retire or quit farming in the next five years. More older farmers planned to sell some land and to reduce machinery inventory. Both of these plans are consistent with plans to retire or quit.

Among the younger farmers, the actual percentage responses differ from the overall average responses, but these differences only increase the relative importance of those areas noted earlier. For example, 88 percent of the younger farmers planned to pay closer attention to marketing, 80 percent planned to keep more complete records, and about 66 percent planned to reduce debt.

Also important among younger farmers are plans to share labor or machinery with neighbors. More younger farmers, as expected, planned to expand by renting or buying additional land.

Table 5.4 shows the percentage of farmers expressing a high or very

TABLE 5.3 Percentages of Sample Farmers Planning to Make Selected Changes in the Near Future, by Operator's Age and Subregion

	Corn Belt		Plains States		Lakes States		NC Region	
	<35	≥60	<35	≥60	<35	≥60	<35	≥60
(Number of Farmers)	(260)	(693)	(134)	(312)	(169)	(304)	(564)	(1,308)
Percentage of Farmers	12.3	32.8	14.2	32.9	16.4	29.6	13.8	32.0
Diversification of Farm Enterprise								
By adding new crops	18.7	7.9	23.2	10.3	27.4	11.1	22.5	9.3
By raising livestock	34.1	22.0	45.8	19.8	33.5	14.5	37.0	19.3
Control of Land								
Buy additional land	30.1	4.6	37.3	5.8	28.7	2.0	31.9	4.3
Rent more acres	42.3	5.2	53.5	7.0	31.7	5.4	42.0	5.6
Rent fewer acres	12.2	13.7	18.3	14.3	14.6	11.8	14.5	13.4
Sell some land	4.9	4.3	3.5	4.9	3.0	8.1	4.3	5.3
Transfer land to lender	1.6	0.9	1.4	0.9	2.4	0.7	2.2	0.9
Operator's Involvement in Farming								
Seek training for new vocation	15.4	2.1	11.3	3.6	15.9	4.1	14.5	3.1
Seek off-farm employment	29.3	11.6	31.0	9.7	34.8	9.8	31.9	10.6
Start a new business (not farming)	12.2	3.0	13.4	3.3	12.8	4.4	13.0	3.4
Retire from farming	3.3	28.0	1.4	27.7	3.0	32.4	2.9	28.9
Quit farming	4.1	18.3	2.1	17.9	4.3	19.9	3.6	18.3

(*continues*)

TABLE 5.3 (continued)

	Corn Belt		Plains States		Lakes States		NC Region	
	<35	≥60	<35	≥60	<35	≥60	<35	≥60
Reduction of Risk Exposure								
Reduce long-term debt	63.4	39.0	65.5	42.2	66.5	35.5	65.2	38.8
Reduce short-term debt	65.9	34.5	68.3	37.7	68.3	34.8	67.4	35.1
Use the futures market to hedge prices	30.1	7.9	31.0	10.0	26.8	7.8	29.0	8.4
Change from cash rent to crop share	19.5	6.1	16.2	9.4	7.9	5.4	15.2	6.5
Buy crop insurance	40.7	22.9	52.8	24.6	45.1	29.4	45.7	24.8
Reduction of Cash Expenditures								
Postpone major farm purchase(s)	43.1	34.8	49.3	34.7	55.5	31.8	49.3	33.5
Share labor or machinery with neighbors	47.2	25.9	38.0	26.7	50.6	22.3	46.4	24.8
Reduce expenditures for hired help	30.9	27.4	28.9	25.2	42.1	19.9	34.1	24.8
Reduce machinery inventory	13.8	19.2	17.6	19.8	16.5	16.6	15.9	18.6
Financial Management and Marketing								
Keep more complete financial records	75.6	41.5	81.0	40.4	85.4	39.2	80.4	40.4
Pay closer attention to marketing	83.7	50.9	95.1	53.2	89.0	45.3	88.4	49.0

TABLE 5.4 Percentages of Sample Farmers Reporting a Need for Information and Training, by Operator's Age and Subregion

	Corn Belt		Plains States		Lakes States		NC Region	
	<35	≥60	<35	≥60	<35	≥60	<35	≥60
(Number of Farmers)	(260)	(693)	(134)	(312)	(169)	(304)	(564)	(1,308)
Percentage of Farmers	12.3	32.8	14.2	32.9	16.4	29.6	13.8	32.0
Production Technology								
Reducing production costs through low-input farming methods	78.9	52.4	81.7	62.3	79.9	50.3	79.7	54.0
Using new technologies as they become available	82.9	53.4	94.4	59.9	84.8	49.3	86.2	53.7
Using appropriate conservation techniques	65.0	43.9	71.8	49.8	62.8	41.6	65.9	44.4
Business Management and Marketing								
Available government assistance	61.8	42.7	67.6	51.4	62.2	39.9	63.8	44.1
Marketing skills	82.1	50.0	89.4	52.9	79.9	43.6	83.3	49.1
Bookkeeping and financial systems	67.5	32.3	80.3	38.6	67.1	28.4	70.3	32.9
New Inputs and Enterprises								
Using new machines and chemical inputs to increase production	79.7	46.3	79.6	54.7	76.8	42.2	79.0	47.2
Diversification of farm operation by adopting new crops and livestock	61.0	33.2	74.6	35.0	65.2	30.1	66.7	32.6
On-Farm Processing								
Processing farm products on farm before selling	42.3	21.3	39.4	20.4	39.6	16.6	40.6	19.9

high need for information and training in selected areas, as reported for two age groups and the three subregions. Younger farmers expressed more need for information and training in all the suggested items in production technology, business management and marketing, use of new inputs and enterprises, and on-farm processing of farm products. In regard to plans for the future and perceived needs for education and training, differences among the three regions and between the two age groups are consistent with the findings we reported earlier.

Responses by Category of Gross Farm Sales

The percentages of farmers planning to make selected changes in the next five years are reported in Table 5.5 for two diverse levels of gross farm sales and the three subregions. Two categories based on gross sales in 1988 are reported from each end of the distribution: very small farms (less than $10,000 in gross sales) and larger farms (sales of $100,000 or more). As stated earlier, about half of the total sample fell into one of these two classes. As expected, more operators of larger farms than of very small farms expected to remain in farming and to rely on the farm as a major source of income.

The larger farms were somewhat more likely to plan to diversify by adding crops or livestock. We found a greater contrast among the three subregions, however, in the extent to which farms of different sizes planned to diversify.

Larger farms were more likely to plan to expand the land base by buying or renting additional land. In contrast, operators of small farms were more likely to plan to seek training for a new vocation or to seek off-farm employment. Differences among the regions are probably due to the availability of off-farm employment. The smaller farmers also were more likely to plan to retire or quit farming. In general, we found more potential change in the operator's involvement in farming among smaller than among larger farmers.

Larger farmers were more likely to plan to reduce long- and short-term debt, to use the futures market, and to buy crop insurance, again with regional differences. Reducing debt, however, may be a higher priority for larger farmers because their actual debt is higher. They also were slightly more disposed to postpone major farm purchases. We noted little difference between the two extremes of size in their plans for sharing labor or machinery with neighbors, reducing expenditures for hired help, and reducing the machinery inventory.

A larger percentage of the operators of larger farms planned to keep more complete financial records and to pay closer attention to marketing.

TABLE 5.5 Percentages of Sample Farmers Planning to Make Selected Changes in the Near Future, by Gross Farm Sales ($1,000) and Subregion

	Corn Belt		Plains States		Lakes States		NC Region	
	<10	≥100	<10	≥100	<10	≥100	<10	≥100
(Number of Farmers)	(475)	(558)	(90)	(368)	(219)	(294)	(781)	(1,222)
Percentage of Farmers	22.5	26.4	9.5	38.9	21.3	28.6	19.1	29.9
Diversification of Farm Enterprise								
By adding new crops	9.8	11.4	16.8	17.7	14.1	17.8	11.5	15.4
By raising livestock	25.3	25.4	22.1	29.3	16.4	22.0	22.0	25.8
Control of Land								
Buy additional land	8.0	22.0	8.4	23.7	6.6	15.7	7.3	21.1
Rent more acres	7.6	37.9	9.5	27.0	3.3	28.0	6.3	32.1
Rent fewer acres	13.3	12.1	16.8	11.1	13.1	15.0	13.6	12.7
Sell some land	3.1	5.3	5.3	5.9	8.0	5.2	4.7	5.7
Transfer land to lender	1.3	1.5	4.2	1.3	0.0	2.1	1.0	1.7
Operator's Involvement in Farming								
Seek training for new vocation	12.0	6.1	11.6	6.4	9.9	7.3	11.0	6.4
Seek off-farm employment	30.2	12.9	30.5	13.4	35.7	14.0	31.9	13.4
Start a new business (not farming)	8.4	7.2	6.3	8.5	11.3	8.4	8.9	8.0

(*continues*)

TABLE 5.5 (continued)

	Corn Belt		Plains States		Lakes States		NC Region	
	<10	≥100	<10	≥100	<10	≥100	<10	≥100
Retire from farming	14.7	8.3	10.5	12.9	17.8	7.3	15.2	9.4
Quit farming	12.9	3.4	9.5	8.7	15.5	5.9	13.6	5.7
Reduction of Risk Exposure								
Reduce long-term debt	40.9	65.5	34.7	65.0	39.0	60.1	39.3	64.7
Reduce short-term debt	37.8	59.1	35.8	62.5	34.7	61.9	36.6	61.2
Use the futures market to hedge prices	5.8	31.1	8.4	27.2	8.5	23.1	6.8	28.1
Change from cash rent to crop share	10.2	11.0	9.5	11.3	5.2	4.5	8.4	9.4
Buy crop insurance	9.8	36.7	25.3	42.9	16.4	46.5	13.6	41.5
Reduction of Cash Expenditures								
Postpone major farm purchase(s)	37.3	44.7	34.7	46.0	34.7	48.3	36.1	46.2
Share labor or machinery with neighbors	35.6	35.2	29.5	35.0	30.5	36.7	33.0	35.8
Reduce expenditures for hired help	24.9	32.2	22.1	27.0	25.4	32.5	24.6	30.8
Reduce machinery inventory	17.8	17.4	20.0	16.5	19.7	15.4	18.8	16.7
Financial Management and Marketing								
Keep more complete financial records	44.9	65.9	49.5	63.8	42.3	66.4	44.5	65.9
Pay closer attention to marketing	47.1	79.9	48.4	82.3	46.5	76.6	46.6	79.9

Table 5.6 presents the percentages of farmers expressing a high or very high need for information and training in selected areas, by two gross farm sales groups and by the three subregions. The operators of the larger farms expressed a need for more information and training in all of the items that were presented for their consideration. A relatively large proportion were interested in using new technologies as they became available, reducing production costs through low-input farming methods, improving their marketing skills, and using new machines and chemicals to increase production.

Summary and Conclusions

Farming is a complex activity requiring skills in agricultural production, marketing, farm finances, and business management. Mastering these skills is important to the economic well-being of the farm household and to society. Correct anticipation of the changes that farmers plan to make in their farm businesses and their perceptions of their needs for education and training would increase the effectiveness of extension and other agricultural education programs and would influence the business decisions of farm input suppliers and farm product processors. In this chapter we contribute to that goal by reporting the respondents' needs for training and education in farm production technology, business management, marketing, use of new farm inputs, and managing new farm enterprises.

The 12-state study area contains considerable differences in climate, natural resources, and urbanization. Regardless of where farm operators live, however, differences in their ages and in the size of their farm businesses affect their plans for the future and their perceptions of education and training needs. We described some of those differences in this chapter; in many cases we simply documented what perceptive persons have noted in their work with farm households. Selected findings are as follows:

1. Two-thirds of the sample farmers planned to pay more attention to marketing in the future. Well over half planned to keep more complete financial records. Although smaller percentages reported a need for training in those areas, this finding supports the need for the traditional extension education areas of marketing and farm business records.
2. Half of all farmers, and two-thirds of those under age 35, planned to reduce their long- and short-term debt. About half of the younger farmers planned to postpone some major purchases. This finding

TABLE 5.6 Percentages of Sample Farmers Reporting a Need for Farming Information and Training, by Gross Farm Sales ($1,000) and Subregion

	Corn Belt		Plains States		Lakes States		NC Region	
	<10	≥100	<10	≥100	<10	≥100	<10	≥100
(Number of Farmers)	(475)	(558)	(90)	(368)	(219)	(294)	(781)	(1,222)
Percentage of Farmers	22.5	26.4	9.5	38.9	21.3	28.6	19.1	29.9
Production Technology								
Reducing production costs through low-input farming methods	42.7	83.3	46.3	83.3	49.8	78.7	45.0	82.3
Using new technologies as they become available	45.8	87.1	48.4	85.9	47.4	89.2	46.6	87.3
Using appropriate conservation techniques	46.7	62.5	48.4	68.1	39.9	65.0	45.0	65.2
Business Management and Marketing								
Available government assistance	33.3	62.9	42.1	66.8	35.2	61.5	35.1	63.9
Marketing skills	43.6	83.3	52.6	82.3	40.8	80.8	44.0	82.6
Bookkeeping and financial systems	34.2	62.9	40.0	62.0	29.6	63.3	33.5	62.5
New Inputs and Enterprises								
Using new machines and chemical inputs to increase production	36.0	82.6	35.8	78.9	41.8	78.3	37.7	80.6
Diversification of farm operation by adopting new crops and livestock	34.7	53.4	42.1	54.2	36.2	50.7	36.1	53.2
On-Farm Processing								
Processing farm products on farm before selling	21.3	33.3	27.4	30.6	22.1	31.8	22.0	32.1

may reflect an attitude toward risk acquired in the farm financial crisis of the last decade, when asset values and farm income declined and caused severe financial stress for leveraged farmers. That model, however, may be inappropriate for many farm businesses today. This observation reinforces the need for educational programs in farm financial management and ineffective use of farm credit.
3. Plans to diversify the farm business by adding crops or livestock varied among regions and were more common among younger farmers. Small and large farms did not appear to differ, however, in their plans to use this strategy.
4. Plans to control more land by purchase or renting also varied among the regions. Younger farmers and larger operators were more likely to plan to expand their farm businesses in these ways.
5. Younger farmers and operators of smaller farms were more likely to have plans to seek training for a new occupation or to seek off-farm employment. The latter option was less likely to be mentioned in the plains states, where non-farm employment opportunities are less plentiful. Younger farmers also were more likely to report plans for starting a new, non-farm business in the near future.
6. Farmers age 60 and older, and those on smaller farms, were more likely to report plans to retire or quit farming in the next five years. The percentage with firm retirement plans, however, was low by non-farm standards: that is, fewer than 30 percent planned to retire within that period.
7. Regarding perceptions of needs for training and education, we noted some regional differences: for example, greater interest in understanding government programs in the plains states, less interest in training in marketing skills in the lakes states, and, in the corn belt, less interest in training in diversifying farm operations by adopting new crops or livestock. Yet even where we found "less interest" in a topic in a particular subregion, a substantial number of farmers may desire that kind of education or training.
8. In keeping with their fewer years of farming experience and the greater number of years in which to recover the costs of an investment in training, a much larger percentage of younger farmers expressed interest in training in all the topics presented. Similarly, a greater percentage of respondents from the larger farms expressed interest in each of the training topics.

Notes

1. Research was supported in part by the Agricultural Experiment Station, College of Agriculture, University of Minnesota-Twin Cities; the Research Division, College of Agricultural and Life Sciences, University of Wisconsin-Madison; and the North Central Regional Center for Rural Development. The assistance of Jackie Fellows in data analysis is acknowledged with thanks.

PART TWO

The Farm Household

Introduction
Katherine Meyer

Restructuring of the agricultural sector of the Midwest economy affected the farm household as well as the farm enterprise. As household members faced the possibility of deteriorating finances and declining ability to consume, they adjusted in multiple ways. These adjustments have been documented in case study research on work, the quality of life, household patterns, and general survival strategies in rural areas (Barlett 1993; Bultena, Lasley and Geller 1986; Fink 1992). Attention has also been paid to the social-psychological effects of rapid economic change (Lobao and Meyer 1991; Armstrong and Schulman 1990). Of course, similar research has been done in non-farm settings, where changing economic conditions have individual level effects on work, life and the psyche.

The research presented here goes beyond case studies to examine the entire Midwestern region, and it extends some hypotheses tested on non-rural populations to a rural one. Yet, we contend that the consequences of economic restructuring are qualitatively different for farm and non-farm families. The family farm is both a production and consumption unit. It is intricately interwoven with the farm enterprise and the rural community in which it resides. Economic changes impact it in a variety of overlapping ways not found in settings where occupation and the labor market are more distinct from the household.

Chapter 6 presents the changes in farm tasks by farm operators and spouses over a five year period. It looks at off-farm employment and changes in decision-making responsibility and the division of labor by gender. As Wozniak, Draugh, and Perch (1988) pointed out, the occupation of farming distinguishes rural families from their non-rural counterparts. Bokemeier and Garkovich (1987) and Lyson (1987) noted that deteriorating finances affected labor allocation for both the farm operation and the household. Here, Fellows examined how off-farm labor fits in with this general picture of declining farm economic prosperity and the blurred distinctions between farm enterprise and household which characterize farming.

Chapter 7 examines how farm families adapted to economic hardship and how these adaptations affected the well-being of farm spouses. The quality of life and adjustments made by households are considered in the context of different economic indicators. A broader picture of household adaptation is drawn in light of household socioeconomic characteristics. In his analysis, Lasley follows up on the observations by Heffernan and Heffernan (1986) and Rosenblatt and Anderson (1981) who postulated linkages among farm family well-being, the adjustments made by farm families, and the variations in economic difficulty they face. Lasley explored those proposed linkages in our Midwest population and systematically examined them with empirical findings.

Chapter 8 centers on the social-psychological effects of the restructuring of the rural Midwest on farm operators and spouses, especially women. It builds on work done by others (Davis and Salamon 1988; Lobao and Meyer 1990; Whatmore 1991) and extends it to include analysis of coping strategies used by rural men and women. Social-psychological research on stress and depression frequently documents how different styles of coping moderate economic and social hardship more or less effectively. The moderation is manifested in varying levels of stress and depression. Meyer brings this concern with coping strategies to the research on stress, depression, and gender in rural populations.

6

The Changing Division of Labor on Family Farms

Jackie Fellows and Paul Lasley

Current research highlights the importance of the division of labor among farm families. The involvement of both spouses or partners, in the case of non-married couples, is required in managing and operating the farm as a successful business. The family farm, however, is not only the place of business for family members; it is also their home. Thus, work and family roles are closely interrelated (Deseran and Simpkins 1991).

Because of this relationship between work and family roles, a farm family's division of labor is influenced both by the composition of the household and by the labor needs of the farming operation. The allocation of family labor is organized to meet both of these needs (Bokemeier and Garkovich 1987). This close meshing of family and work roles distinguishes farm families from most non-farm families (Wozniak, Draughn, and Perch 1988). Changes within the farm business hold important implications for changes in labor allocation both for the farm operation and for household maintenance (Bokemeier and Garkovich 1987; Deseran, Falk, and Jenkins 1984; Lyson 1985; Wozniak et al. 1988).

The influence of declining farm prices and depreciated asset values on the farm family's division of labor has been documented by numerous studies conducted during the farm crisis of the 1980s. During that decade, many farm families altered their division of labor by one or both spouses taking off-farm employment in an effort to ensure the survival of the family farm (Bartlett 1986; Deseran 1985; Lyson 1985; Ollenburger, Grana, and Moor 1989). Taking an off-farm job to earn supplemental income, improve cash flow, or to qualify for medical insurance coverage holds important consequences for family dynamics and work roles.

Although economic need is a primary factor for combining off-farm employment with farming, other factors must be considered before a family alters its labor allocation. Previous research (e.g., Bokemeier and Garkovich 1987; Wozniak and Scholl 1988) examining the off-farm labor involvement of farm operators and spouses has found that personal resources and family characteristics, such as the number of dependent children and whether one of the spouses already work off the farm are important considerations in deciding which spouse seeks off-farm work.

In examining personal resources, studies generally have focused on the operators' off-farm employment and have found that education was the strongest predictor. Operators with higher levels of education were more likely to take off-farm employment (Deseran et al. 1984; Sumner 1982; Wozniak et al. 1988). In addition, higher levels of education were also found to be a strong predictor of a spouse's off-farm employment (Rosenfeld 1986; Sweet 1972; Wozniak et al. 1988; Sumner 1982).

In previous studies examining the effects of family composition (the number of children in a family) on off-farm employment patterns of operators and spouses, there are mixed findings. Sumner (1982) reported that the number of children in a family had little effect on farm operators off-farm labor patterns. However, Deseran et al. (1984) found that the number of children in a family had a significant effect on spouses' off-farm earnings, while exerting little or no effect on the decisions of operators to take off-farm jobs.

Off-farm employment of the husband also has been shown to affect the wife's involvement in the daily farm operation and in making decisions about the farm. When the operator works off the farm, the spouse is likely to have more influence in farm decisions (Deseran and Simpkins 1991) and to be more involved in the daily farm operation (Lyson 1985). Boulding (1980) and Rosenfeld (1986), however, note that the relationship between a woman's involvement in farm tasks and her input into farm decisions is not always clear cut. In some instances, the wife may provide much of the needed labor, but defer to her husband for decisions, yet in other cases there may be a closer correspondence.

Most of these studies are limited in several respects. With the exception of the multi-state study conducted by Wozniak and her colleagues (1988), the studies tend to focus on one state or community; thus it is difficult to generalize the findings to other areas of the country. Few studies have examined the off-farm labor patterns of Midwestern farm families, the area most adversely affected by the farm crisis.

In addition, only a few studies have included both the operator and the spouse in examining off-farm labor patterns and, generally, have not addressed the relationship between off-farm and on-farm labor arrangements within the family (Simpson, Wilson, and Young 1988). This has greatly limited our understanding of the changing nature of family farming. This chapter seeks to identify and explicate the personal, familial, and farm characteristics that are related to farm families' division of labor.

Data and Methods

Families were placed into categories based upon their off-farm employment status: no off-farm employment, spouse has off-farm employment, operator has off-farm employment, or both are employed in off-farm jobs.

To classify families by net income, three categories were used: less than $30,000; $30,000 to $59,999; and $60,000 or more. Families were categorized by the number of hours of work per week respondents reported: fewer than 30 hours, 30 to 60 hours, and more than 60 hours. The number of hours worked per week in an off-farm job was divided into four categories because some families did not report off-farm employment: no off-farm work, fewer than 30 hours, 30 to 60 hours, and more than 60 hours.

Respondents were also classified by the number of years of off-farm employment, the number of weeks of off-farm employment in 1988, the miles they commuted to work, and the occupation held by the family member(s). Years of off-farm employment were divided as follows: 1 to 5 years, 6 to 10 years, 11 to 20 years, and more than 20 years. Weeks worked in off-farm jobs in 1988 were: 1 to 13 weeks, 14 to 26 weeks, 27 to 39 weeks, and 40 to 52 weeks. Eight categories of occupation were included: professional, managerial/administrative, sales/clerical, craftsman/machine operator, transportation, laborer/service worker, self-employed, and other.

In addition to these variables, age, education, and number of children were included in the analysis. In addition, three additional sets of questions to ascertain the work roles of women were analyzed. These questions were: how frequently spouses engaged in certain farm and household tasks (always, sometimes, never/not done); whether the time they spent on these tasks had increased, decreased, or remained the same from 1984 to 1988; and who was involved in making decisions regarding land, the purchase of household appliances and major farm equipment, and yearly farm operation decisions.

Findings

Total Sample

Socio-demographic characteristics. The average age of the farm operators was 52 years compared to 49 years for the spouses. Twelve years of education was the average for operators, while spouses had an average of 13 years of education. Farm families in this study had an average of two children.

Approximately six out of ten families (63 percent) had a net family income of less than $30,000. A loss was reported by 5 percent; 7 percent reported a net family income of over $60,000.

In six out of ten families, one or both adult members had taken off-farm employment in 1988. In 31 percent of the families, one member was working in an off-farm occupation; 29 percent responded that both adult members had off-farm occupations.

Decision-making responsibilities. Farm families are faced with major decisions concerning both household and farm operations. The findings show a definite pattern in regard to the person or persons most likely to make these decisions. A joint decision by the operator and the spouse was most likely when the decision concerned an investment such as buying a major household appliance (76 percent), expanding or contracting the farm operation by buying or selling land (61 percent), or renting more or fewer acres (46 percent). However, the operator was more likely to be responsible for decisions about farm business operations such as what crop or livestock to produce (47 percent), when to sell the farm's products (53 percent), and whether to try a new agricultural practice (54 percent).

Family division of labor. In most of the families, the division of labor followed traditional gender lines for on-farm work. Three-fourths of the operators worked on the farm more than 30 hours a week, while 63 percent of the spouses responded that they worked fewer than 30 hours a week on the farm. Comparing off-farm work patterns: 30 percent of the operators worked 30 hours or more at an off-farm job, compared with 28 percent of the spouses reported working off the farm 30 hours or more.

Farm and household tasks performed by spouses. Spouses were involved in many of the farm tasks as well as in household maintenance, and in some cases, in working off the farm (Table 6.1). In

TABLE 6.1. Farm Tasks Performed by Spouse and Change in Time Spent Performing Tasks in Past Five Years

	Tasks			Time Spent		
	Always	Sometimes	Never/ Not Done	Increased	Same	Decreased
	---percent---					
Did field work	10	51	39	14	55	31
Milked/cared for animals	20	42	38	15	54	31
Ran farm errands	29	63	8	17	66	17
Purchased major farm supplies and equipment	5	25	70	5	83	13
Marketing farm products through wholesale buyers or directly to consumers	5	19	76	6	81	13
Kept books/ maintained records	50	28	22	24	65	11
Did household tasks and/or child care	92	6	2	20	66	14
Supervised farm work of others	4	35	61	7	77	16
Took care of a vegetable garden or animals for family consumption	57	30	13	12	67	21

50 percent or more of the families, caring for a garden or animals for family consumption (57 percent) and/or bookkeeping and maintaining records (50 percent) were always the spouse's tasks.

Many of the spouses also reported that they sometimes helped with a variety of farm tasks. Sixty-three percent sometimes ran farm errands, 51 percent sometimes did field work, and, if the farm raised animals for the market or was a dairy producer, 42 percent sometimes milked or cared for those animals.

Spouses, however, were less likely to be involved in purchasing major supplies and equipment, in marketing products, or in supervising the farm work of others. Purchasing supplies and equipment was never done by 70 percent of the spouses, 76 percent stated that they never marketed the farm's products, and 61 percent never were involved in supervising any farm work by other people.

Time spent on tasks. For most spouses, the time spent on farm tasks had remained the same over the past five years. For nearly one-fourth of the spouses (24 percent), however, the time spent on bookkeeping and maintaining records had increased, while for 31 percent, the time spent on field work and milking and/or caring for animals had decreased.

Off-farm employment. For operators, the largest proportion (37 percent) reported having been employed off the farm for more than 20 years suggesting that they are permanent part-time farmers. In contrast, the largest proportion of the spouses (37 percent) stated that they had taken off-farm employment in the last five years. More than one-fourth of the operators and the spouses (26 percent) worked off the farm for 11 to 20 years.

Operators commuted longer distances to their jobs than their spouses. Thirty-five percent of the operators reported driving over 20 miles one way to work; only 15 percent of the spouses reported driving such a distance. Operators were more likely to report that they had labor or service occupations (23 percent) or worked as craftsmen or machine operators (19 percent). Spouses, on the other hand, were more likely to report a professional occupation (29 percent), including nursing and teaching or to be employed in sales or clerical work (27 percent).

Comparisons Between Families with Different Employment Status

Socio-demographic characteristics. One-way Anovas comparing the employment groups by age, education, and number of children indicated significant differences between the groups for operators and spouses age, educational level, and the average number of children in the household (Table 6.2). A Scheffe procedure (p < .05) to determine which groups differed revealed that families with no off-farm work were significantly older than families in all three employment groups. The average age of operators in families with no off-farm work was 57 years and among spouses the average was 54 years.

Operators and spouses in families in which the operator worked off the farm tended to be older than operators and spouses in which the spouse or both adult members worked off the farm. In families in which the operator had off-farm employment, the average age of the operator was 51 years and for the spouse, the average age was 49 years. In households in which the spouse had off-farm employment, the average age for the operator was 48 years and 45 years was the average age of the spouse. When both adult members worked, the average ages for operator and spouse was 47 years and 44 years.

Significant differences exist in levels of education between employment statuses. Operators in households which reported no off-farm employment had fewest years of education. Operators had an average of 11.7 years of education and the spouses, on average, had 12.1 years of schooling. In families in which the operator worked off the

TABLE 6.2 Comparison Between Families with Different Employment Status

	No Off-Farm Employment	Spouse Works	Operator Works	Both Work	
	----------percent----------				
Age (years)					
Operator	57	48	51	47	F=166.63 *
Spouse	54	45	49	44	F=175.61 *
Education (years)					
Operator	11.7	12.7	12.0	13.0	F=53.01 *
Spouse	12.1	13.4	12.0	13.4	F=75.78 *
Number of Children	2.3	2.1	2.5	2.1	F=6.61 *
Proportion with Income Less Than $30,000	68	65	56	51	*
Proportion with Income Greater Than $80,000	7	7	9	9	NS

* = p ≤ .01

farm, the average was 12 years for both the operator and the spouse. When the spouse worked off the farm, operators had an average of 12.7 years and the spouses averaged 13.4 years of schooling. In households where the operator and the spouse were employed off the farm, operators averaged 13 years and spouses 13.4 years of education.

No significant differences in number of children were found between families with no off-farm employment and the three off-farm employment groups. The Scheffe test showed a significant difference in number of children between families in which the operator had off-farm employment and families in which the spouse or both adult members had off-farm employment. Families with no off-farm income had an average of 2.3 children; those families with the spouse or both adult members working off the farm had an average of 2.1 children, and families with the operator employed off the farm had an average of 2.5 children.

Although most of the families in all groups reported less than $30,000 a year, a higher percentage of families with no off-farm employment (68 percent) belonged to this income category. When the spouse was employed, 65 percent made less than $30,000 in 1988; when the operator was employed, 56 percent were in this category. Fifty-one percent of the households with both members employed made less than $30,000.

Among the four employment groups, the largest percentage (7 percent) of families reporting a loss of net income in 1988 were families in which the spouse worked. Households with no off-farm employment had the smallest percentage reporting a loss (4 percent). Five percent of the households in which the operator worked off the farm reported a loss in 1988 and 6 percent of the households in which both the operator and the spouse had off-farm employment had a net income loss.

For families in which the operator worked or in which both members worked, 9 percent of each group reported an income of $60,000 or more. This income level was reported by 7 percent of the families with no off-farm employment and of families in which the spouse worked off the farm.

Employment status affected the number of hours the operator or the spouse worked on the farm. Among operators who indicated that they worked fewer than 30 hours a week on the farm, the smallest percentage (8 percent) was found when the spouse was employed. In households with no off-farm employment, 20 percent of the operators reported working fewer than 30 hours a week on the farm. When the operator was employed or when both adult members worked, however, more than 50 percent of the operators reported working fewer than 30 hours a week on the farm (operators, 54 percent; both working, 55 percent).

In families with no off-farm employment or in which the spouse worked, the operator was more likely to work more than 60 hours a week on the farm. More than one-third of the operators in these two groups worked more than 60 hours (no off-farm employment, 36 percent; spouse working, 38 percent). Only 10 percent of the operators who worked off the farm and 12 percent of the operators in families with both adult members working off the farm reported working more than 60 hours on the farm.

The spouse's employment status also affected how many hours he or she worked on the farm. When spouses were employed, they were more likely to report working fewer than 30 hours on the farm (spouses, 72 percent; both working, 77 percent). When the operator was employed off the farm, 63 percent of the spouses worked fewer than 30 hours a week on the farm. In households with no off-farm employment, 48 percent of the spouses worked fewer than 30 hours.

The highest percentage of spouses (14 percent) worked more than 60 hours a week in households in which there was no off-farm employment. When the family had off-farm employment, 5 percent or fewer of the spouses reported working on the farm more than 60 hours (spouse works, 5 percent; operator works, 4 percent; both work, 3 percent).

Decision-making responsibilities. Table 6.3 shows the different patterns of decision making in farm families, depending upon the decision, and who, if anyone, was employed off the farm. A Chi-Square was used to determine if there were significant differences between the four employment groups. In comparing the differing employment groups, no significant differences were found when deciding whether to buy or sell land, to buy major household appliances, or to buy major farm equipment. But the families in the four employment groups differed significantly in their decision-making patterns when deciding to produce a new crop or livestock ($X^2=21.88$, df=9, $p < .01$), when to sell their agricultural products ($X^2=30.89$, df=9, $p < .00$), whether to try a new agricultural practice ($X^2=30.30$, df=9, $p < .00$), and whether to rent more or less land ($X^2=25.58$, df=9, $p < .00$).

Although the spouse was less likely to be involved in the decision to produce a new crop or livestock in all four employment groups, a larger percentage of families reported that this was the operator's decision when the spouse was employed off the farm (54 percent) or both adult members had off-farm employment (46 percent). The operator was the sole decision maker in 44 percent of the households that reported no off-farm employment, while 42 percent of the households in which the operator worked off the farm indicated the operator made this decision. In contrast, a joint decision to rent more or

TABLE 6.3 Relationship Between Decision-Making Responsibility and Off-Farm Employment as Reported by Spouse

	No Off-Farm Employment	Spouse Works	Operator Works	Both Work
	------------------percent------------------			
Buy or sell land				
Usually myself	1	1	1	1
My spouse or someone else	18	19	17	15
Myself and spouse or someone else	59	60	58	63
Decision has never come up	22	20	24	21
Rent more or less land				
=Usually myself	1	--	1	1**
My spouse or someone else	26	34	24	29
Myself and spouse or someone else	48	46	45	43
Decision has never come up	25	20	30	27
Buy major household appliances				
Usually myself	13	16	15	11
My spouse or someone else	8	8	8	8
Myself and spouse or someone else	77	75	76	80
Decision has never come up	2	1	1	1
Buy major farm equipment				
Usually myself	2	1	1	1
My spouse or someone else	44	51	42	47
Myself and spouse or someone else	48	44	51	45
Decision has never come up	6	4	6	7
Produce a new crop or livestock				
Usually myself	1	1	2	1*
My spouse or someone else	44	54	42	46
Myself and spouse or someone else	35	30	37	34
Decision has never come up	20	15	19	19
Decide when to sell your agricultural products				
Usually myself	2	2	3	2**
My spouse or someone else	49	62	48	55
Myself and spouse or someone else	41	31	39	35
Decision has never come up	8	5	10	8
Decide to try a new agricultural practice				
Usually myself	1	1	1	1**
My spouse or someone else	53	63	50	57
Myself and spouse or someone else	30	25	34	25
Decision has never come up	16	11	15	17

-- Less than 1 percent
* $p < .01$
** $p < .00$

less land was more likely to be made in households in which the operator had off-farm employment (37 percent) and less likely to be a joint decision when the spouse was employed off the farm (30 percent).

In all employment groups, the operator was more likely to decide when to sell the family's agricultural products. However, depending on the family's employment status, there were significant differences in the percentage of families reporting that this was the operator's decision. More than one-half of the families stated this was the operator's decision when the spouse (62 percent) or both adult members (55 percent) had off-farm employment. In contrast, the operators decided when to sell their agricultural products in 49 percent of the families who reported no off-farm employment and 48 percent of the families in which the operators had off-farm employment.

This same pattern of decision making was used by families in the four employment groups when the decision was to try a new agricultural practice. The highest percentage of families who reported that this was the operator's decision were families in which the spouse had off-farm employment (63 percent) or both adult members worked off the farm (57 percent). In families with no off-farm employment, 53 percent indicated the operator made this decision and, again, the lowest percentage of families reporting this was the operator's decision were families in which the operator had off-farm employment (50 percent).

When the decision was jointly made, the percentage was higher if the spouse was not working off the farm. Thirty-four percent of the families in which the operator worked off the farm reported this was a joint decision and 30 percent of the families with no off-farm employment stated this was a joint decision. In contrast, only 25 percent of the families in which the spouse or both adult members had off-farm employment indicated this was a joint decision.

A slightly different pattern of decision making is observed in regard to renting more or less land. This was a joint decision by 48 percent of all the families. The highest percentages reporting they jointly decided to rent more or less land were families in which there was no off-farm employment (48 percent) or in which the spouse was employed (46 percent). Forty-five percent of the families in which the operator worked and 43 percent of the families with both adult members working off the farm made a joint decision on renting more or less land. But, it should be noted that the highest percentages reporting this decision had not come up were families in which the operator had off-farm employment (30 percent) or both adult members were employed off the farm (27 percent). One-quarter of the families with no off-farm employment and 20 percent of the families in which the spouse had off-farm employment stated they had not had to make that decision.

Spouses' involvement in farm tasks. Table 6.4 shows that most of the spouses were involved in farm tasks in all employment groups. Depending on the spouses' employment status, there were significant differences in how often the spouses were involved in five of the tasks: field work ($X2=23.60$, $df=9$, $p < .00$), milking or caring for farm animals ($X2=55.36$, $df=9$, $p \sim .00$), running farm errands ($X2=34.54$, $df=9$, $p < .00$), purchasing farm supplies and equipment ($X2=21.93$, $df=9$, $p < .01$), and gardening or tending animals for consumption ($X2=51.14$, $df=9$, $p < .00$).

When spouses were not employed off the farm, they were more likely to report they always performed the task. In families with no off-farm employment, 13 percent of the spouses always did field work, 25 percent always milked or cared for farm animals, 33 percent always ran farm errands, 6 percent always purchased farm supplies and equipment, and 62 percent always had a garden or tended animals for family consumption.

When the operator was employed off the farm, 11 percent of the spouses reported always doing field work, 23 percent always milked or cared for farm animals, 30 percent always ran farm errands, 5 percent always purchased farm supplies and equipment, and 63 percent gardened or tended farm animals for family consumption.

In contrast, when the spouse was employed off the farm, 7 percent of the spouses reported always doing farm work, 14 percent always milked or cared for farm animals, 25 percent always ran farm errands, 4 percent always purchased farm supplies and equipment, and 48 percent always gardened or tended animals for family consumption.

In households in which both adult members were employed off the farm, 9 percent of the spouses said they always did field work, 15 percent always milked or cared for farm animals, 25 percent always ran farm errands, 4 percent always purchased farm supplies and equipment, and 50 percent always gardened or tended animals for family consumption.

Although no significant differences were found between the employment groups and marketing farm products, bookkeeping and maintaining records, household tasks and child care, and supervising others' farm work when the spouse or both adult members worked off the farm, a higher percentage of spouses reported they never did the task. Eighty-one percent of the spouses who had off-farm employment said they never marketed their farm products and 78 percent never did this task when both operator and spouse was employed off the farm. In households which reported no off-farm employment, 76 percent of the spouses never marketed farm products and in families in which the operator worked off the farm, 74 percent of the spouses never marketed their farm products. This suggests in families where spouses are

TABLE 6.4 Relationship Between Spouse's Farm Tasks and Off-farm Employment as Reported by Spouse

	No Off-Farm Employment	Spouse Works	Operator Works	Both Work
	----------percent----------			
Did field work				
Always	13	7	11	9**
Sometimes	52	52	51	52
Never/not done	35	41	38	39
Milked or cared for animals				
Always	25	14	23	15**
Sometimes	40	45	36	44
Never/not done	35	41	41	41
Ran farm errands				
Always	33	25	30	25**
Sometimes	59	69	62	68
Never/not done	8	6	8	8
Purchased major farm supplies and equipment				
Always	6	4	5	4*
Sometimes	26	21	32	24
Never/not done	68	75	63	72
Marketed farm products through wholesale buyers or directly to consumers				
Always	5	4	5	5
Sometimes	19	15	21	17
Never/not done	76	81	74	78
Kept books and maintained records				
Always	53	48	48	49
Sometimes	27	27	33	27
Never/not done	20	25	19	24
Did household tasks and/or child care				
Always	91	91	93	94
Sometimes	7	7	6	5
Never/not done	2	2	1	1
Supervised the farm work of others				
Always	5	3	5	4
Sometimes	38	34	35	35
Never/not done	57	63	60	61
Took care of a vegetable garden or animals for family consumption				
Always	62	48	63	50**
Sometimes	27	35	27	37
Never/not done	11	17	10	13

* $p < .01$
** $p < .00$

employed off the farm, that they are generally less involved in farm decision making.

Bookkeeping and maintaining farm records was never done by 25 percent of the spouses in families in which the spouse was employed off the farm and 24 percent of the spouses never did this task in families in which both adult members worked off the farm. Twenty percent of the spouses in households with no off-farm employment and 19 percent of the spouses in families in which the operator worked off the farm never did the paperwork

When the spouse worked off the farm, 63 percent of the spouses never supervised others; 61 percent of the spouses never did this task when both adult members had off-farm employment. A smaller percentage of the spouses in households with no off-farm employment (57 percent) or where the operator worked off the farm (60 percent) never supervised others.

Change in time spent at task. Table 6.5 shows the changes in the time spouses spent on tasks in the past five years. Most families reported no change, but among families who reported changes, the spouses were more likely to report that the time they spent had decreased except on domestic tasks, on the paperwork involved with farming, and (for spouses at home) on running farm errands. A Chi square showed no significant differences between the groups, with one exception, the time spent on milking or caring for farm animals ($X^2=14.84$, $df=9$, $p\sim.02$).

In households with a spouse employed, 33 percent of the spouses reported that the time they spent on animal care had decreased. About three out of ten spouses had decreased their time spent on animal care (28 percent) in households in which both operator and spouse had off-farm employment. Approximately one-third of the spouses in families with no off-farm employment reported that they spent less time caring for animals (31 percent); one-fourth of the spouses with the operator working off the farm had decreased the time they spent on animal care.

Although no significant differences were found when comparing the time spent by spouses on the other tasks, a higher percentage of spouses in families in which the operator had off-farm employment had increased the time they spent doing field work, running farm errands, supervising the work of others, and taking care of a vegetable garden or animals for family consumption. A slightly higher percentage of spouses who had off-farm employment stated the time they spent milking or caring for animals, running farm errands, doing the paperwork involved with farming, and taking care of a vegetable garden or animals for family consumption had decreased.

TABLE 6.5 Relationship Between Change in Time Spent on Tasks in Past Five Years and Off-farm Employment as Reported by Spouse

	No Off-Farm Employment	Spouse Works	Operator Works	Both Work
	----------percent----------			
Did field work				
Increased	13	14	16	14
The same	54	55	61	57
Decreased	33	31	23	29
Milked or cared for animals				
Increased	14	16	16	12*
The same	55	51	59	60
Decreased	31	33	25	28
Ran farm errands				
Increased	17	17	19	17
The same	68	63	67	65
Decreased	15	20	14	18
Purchased major farm supplies and equipment				
Increased	3	6	6	4
The same	83	84	84	85
Decreased	14	10	10	11
Marketed farm products through wholesale buyers or directly to consumers				
Increased	4	7	6	5
The same	81	81	83	83
Decreased	15	12	11	12
Kept books and maintained records				
Increased	26	25	24	22
The same	64	63	68	68
Decreased	10	12	8	10
Did household tasks and/or child care				
Increased	19	19	22	22
The same	67	67	66	65
Decreased	14	14	12	13
Supervised the farm work of others				
Increased	6	8	9	6
The same	78	77	78	78
Decreased	16	15	13	16
Took care of a vegetable garden or animals for family consumption				
Increased	10	12	14	13
The same	68	65	70	66
Decreased	22	23	16	21

* $p < .02$

Conclusion

Full-time farming has become increasingly difficult for farm families in the last 50 years (Danes and Keskinen 1990). Various researchers have noted that in a significant proportion of families, one or both adult members work at an off-farm occupation (i.e., Danes and Keskinen 1990; Deseran 1985; Ollenburger et al. 1989; Zook 1988). To take advantage of this strategy, families have had to alter their division of labor and, in some cases, the responsibility for decision making.

This chapter has focused on the changing division of labor in families in response to the farm crisis. The findings highlight the significance of personal resources in deciding to take off-farm employment and determining who will work off the farm. The results of this study show that younger families with higher levels of education are more likely to choose off-farm employment as a strategy for adapting to economic stress. Young families may not have the resources necessary to try other options, but given that younger families have more years of education, off-farm employment is a more viable option.

The findings also point out the importance of family size in the decision of who would work off the farm. In those families in which one or more of the adult members had off-farm employment, the operator was more likely to work off the farm if the average size of the family was more than two children. This finding supports Sumner's (1982) and Deseran's (1984) findings that the effect of family composition was gender specific. We can find an explanation if we examine the proportion of spouses who responded that they always did the domestic tasks and care in the family. This response strongly suggests that spouses who work off the farm cannot alter the time they spend on those tasks and therefore must change the time they spend on farm labor or must add more hours to their schedule.

The results show that off-farm employment has helped to offset financial stress resulting from income fluctuations during the farm crisis. Altering the family's division of labor to include off-farm employment was financially beneficial for many families, especially those families in which the operator was working off the farm or in which both adult members had off-farm occupations. Because of this alteration in the division of labor, the spouse and/or older children may have taken more responsibility for the farm work.

The pattern of decision making found in this study supports Deseran and Simpkins' finding (1991) that the spouses who remained on the farm were more likely to be involved in farm business decisions. One reason may be that these types of decisions must take into consideration the labor performed by the spouse, who may be responsible for doing the work once the decision is made.

The results also show that spouses who remain at home are more likely to involve themselves in the daily farm operations, thus combining their domestic labor with unpaid labor. The patterns of involvement in either on-farm or off-farm work suggest that the family takes into account each member's personal resources and the labor demands of the farm in creating a division of labor that will maximize each member's contribution of work.

Although this study has a number of limitations, it highlights the importance of understanding the different possible forms of family farming. It also draws attention to the need for more extensive research on how the family adapts its division of labor to changes in the socioeconomic conditions facing the farm sector in the United States.

It is evident that part-time farming shields farm families from the vagaries of the market place and affords income protection and needed benefits such as health insurance. The farm crisis of the 1980s has contributed to increased off-farm employment as a coping strategy. Future research should explore the extent to which decisions made during the 1980s to take off-farm employment become a permanent part of family farming. It is likely that even though many families took off-farm jobs as a temporary solution, these jobs may become a permanent feature for many families.

References

Bartlett, Peggy. 1986. "Part-time Farming: Saving the Farm or Saving the Lifestyle?" *Rural Sociology* 51(3): 289-313.

Bokemeier, Janet and Lorraine Garkovich. 1987. "Assessing the Influence of Farm Women's Self Identity on Task Allocation and Decision-making." *Rural Sociology* 52(1): 13-36.

Boulding, Elise. 1980. "The Labor of U.S. Farm Women." *Sociology of Work and Occupations* 7: 261-290.

Danes, Sharon M. and Susan M. Keskinen. 1990. "The Extent of Off-farm Employment and Its Impact on Farm Women." *Human Services in the Rural Environment* 14(1): 10-14.

Deseran, Forrest A. 1985. "Off-Farm Employment and Social Networks of Louisiana Farm Couples." *Sociologia Ruralis* 25: 174-187.

Deseran, Forrest A., William W. Falk, and Pamela Jenkins. 1984. "Determinants of Earnings of Farm Families in the U.S." *Rural Sociology* 49(2): 210-229.

Deseran, Forrest A. and Neller Ree Simpkins. 1991. "Women's Off-farm Work and Gender Stratification." *Journal of Rural Studies* 7: 91-97.

Lyson, Thomas A. 1985. "Husband and Wife Work Roles and the Organization and Operation of Family Farms." *Journal of Marriage and Family* (August): 759-764.

Ollenburger, Jane C., Sheryl J. Grana, and Helen A. Moor. 1989. "Labor Force Participation of Rural Farm, Rural Nonfarm and Urban Women: A Panel Update." *Rural Sociology* 54(3): 533-550.

Rosenfeld, Rachel. 1986. "U.S. Farm Women: Their Part in the Farm Work and Decision-Making." *Work and Occupations* 13: 179-202.

Simpson, Ida Harper, John Wilson, Kristina Young. 1988. "The Sexual Division of Farm Household Labor: A Replication and Extension." *Rural Sociology* 53(2): 145-165.

Sumner, J. A. 1982. "The Off-Farm Labor Supply of Farmers." *America Economic Review* 64: 499-509.

Sweet, James. 1972. "The Employment of Rural Farm Wives." *Rural Sociology* 37(4): 553-577.

Wozniak, Patricia and Kathleen K. Scholl. 1988. "Employment Decisions of Farm Couples: Full-Time or Part-Time Farming?" *Home Economics Research Journal* 17: 20-35.

Wozniak, Patricia J., Peggy S. Draughn, and Karen L. Perch. 1988. "A Multi-State Study of Off-Farm Employment." *Home Economics Research Journal* 17: 10-19

Zook, Lee. 1988 "Gender Role Changes in Farm Families." *Human Services in the Rural Environment* 11(3): 7-9.

7

The Impacts of Financial Hardship on Familial Well-Being

Paul Lasley

This chapter explores household and individual behavioral changes that farm families made in response to economic hardship, and relates these adaptations to dimensions of familial well-being. People react differently, but in predictable ways, when faced with losing their job, unexpected bills, a sudden loss of income, or financial uncertainty. Reducing household and personal living expenditures, seeking additional work to earn supplemental income, and making lifestyle changes are common responses. These "belt-tightening" measures often trigger social and emotional adjustments. Responses to economic hardship may result in perceived declines in familial well-being, including such elements as diminished quality of life, decreased levels of job satisfaction, and lower levels of community attachment.

Most of the research on the consequences of economic hardship has focused on emotional stress and has drawn largely on non-farm populations (Brenner 1973; Cobb and Kasl 1977). For many farm families, farming as a business is intertwined closely with farming as a way of life. Although recent studies have called attention to the unique features of farm families, little research has examined systematically how economic hardship is reflected in their household adaptations, which in turn affect familial well-being (Heffernan and Heffernan 1986; Rosenblatt and Anderson 1981). The virtues of family farming have long been extolled, but accumulating evidence shows that farm families are no better prepared than the non-farm population to cope with economic hardship. In fact, the close association between work and family life in farm families may cause additional negative consequences. Where the farm operator is viewed as both the owner and the manager of the business, his or her actions and decisions can be observed closely and judged by other family members. This situation may create additional pressure.

Drawing from research on the effects of job loss and unemployment (Banks and Jackson 1982; Cobb and Kasl 1977; Dooley and Catalano 1980), rural sociologists have linked economic hardship in farm families with a number of social and psychological pathologies. Heffernan and Heffernan (1986) found that economic trouble was related to feelings of depression, withdrawal from social interaction in the community with friends and family, and mood swings. Research conducted by Bultena, Lasley, and Geller (1986) found that financial distress influenced household expenditures. In follow-up work, Geller, Bultena, and Lasley (1988) used the life events perspective to further elucidate the relationship between financial stress and psychosocial consequences.

Less well understood is how farm economic hardship influences the sense of familial well-being. Familial well-being is conceptualized as perceptions of quality of life, community attachment, job satisfaction, and commitment to farming. Although it is generally accepted that economic hardship results in interpersonal and familial stress, we lack clear evidence showing how such hardship shapes farm family members' perceptions of their overall familial well-being. Perhaps one of the greatest tragedies among many farm families during the farm crisis was the degree of personal and familial sacrifice they made to save the farm business. Many of these families struggled to keep the farm financially solvent, only to discover that the adjustments they made to save the farm business contributed to the dissolution of the family. Broken marriages, spouse and child abuse, alcohol and drug dependency, loss of self-esteem, and emotional distress were some of the social path-ologies that resulted from economic hardship (Barrett 1987; Prairiefire 1987).

This chapter explores the relationship between economic hardship and selected measures of familial well-being. We predict that financial hardship triggers adjustments in the farm household, which in turn influence perceptions of familial well-being.

Measurement of Familial Well-Being

In the North Central Regional Farm Survey, familial well-being is conceptualized as operators' and spouses' subjective assessments of their families' quality of life, their personal satisfaction with the job of farming, their attachment to community, and their commitment to farming. Debt-asset ratios and net family income are used as objective measures of financial well-being. Household adaptations made in response to economic need are used as a second measure of financial status.

Quality of Life

Respondents were asked how their family's quality of life had changed over the past five years (1984-1988). This period includes what most observers believe were the worst years of the farm crisis. Respondents were asked to indicate their feelings on a five-item scale that ranged from "become much better" to "become much worse." Only 6 percent of the operators and 7 percent of the spouses stated that their families' quality of life had become much better over the past five years. About one-third (31 percent) of both the operators and the spouses said their families' quality of life had become somewhat better. Taken together, nearly four respondents in ten described their families' quality of life as having improved in the past five years. Forty-six percent of the operators and 45 percent of the spouses believed their quality of life had remained the same. Fourteen percent of both the operators and the spouses reported that their quality of life had become "somewhat worse" in the past five years, and 3 percent felt their quality of life had become "much worse." Farm operators and spouses were quite similar in their assessments of their quality of life.

Satisfaction with Farming

The second dimension of familial well-being was operators' and spouses' responses to the question "All things considered, your satisfaction with farming has . . . " The response categories were ranked on a five-item scale that ranged from "become much better" to "become much worse." Operators were somewhat more likely than spouses to report that their satisfaction with farming had improved over the past five years. Twenty-four percent of the operators stated that their satisfaction with farming had become either much better (4 percent) or somewhat better (20 percent) between 1984 and 1988. About one-half (48 percent) said their satisfaction with farming had remained the same, 22 percent reported that their satisfaction with farming had become somewhat worse, and 6 percent stated that it had become much worse. Among the spouses, 19 percent reported their satisfaction had become either much better (4 percent) or somewhat better (15 percent). Fifty-one percent of the spouses indicated no change in their satisfaction with farming in the past five years, and 30 percent said it was either somewhat worse (23 percent) or much worse (7 percent). Although the difference is not statistically significant, spouses were somewhat less likely than operators to indicate improved satisfaction with farming over the past five years.

Attachment to Community

A third dimension of familial well-being was respondents' attachment to their communities. Respondents were asked "Suppose for some reason you had to move away from here. How sorry would you be to leave?" Response categories were "very sorry," "somewhat sorry," "a little sorry," and "not at all sorry." About one-half of both the operators and the spouses said they would be "very sorry" if they had to move from their community—53 percent of the operators and 50 percent of the spouses. Twenty-nine percent of both operators and spouses stated they would be "somewhat sorry" if they had to move. Only 6 percent of the operators and 7 percent of the spouses said they would be "not at all sorry." On the basis of this measure of community attachment, about eight of ten respondents, regardless of gender, would be either "very sorry" or "somewhat sorry" if they had to leave their community.

Commitment to Farming

The final measure of familial well-being was the question "Would you recommend farming to your children or another relative?" Thirty percent of both operators and spouses said they would recommend farming to their children or another relative. Forty-seven percent of the operators stated they would not do so, compared to 39 percent of the spouses. A slightly higher proportion of the spouses than of the operators were uncertain (31 percent compared to 23 percent). About one-third of the respondents reported that their satisfaction with farming had declined in the past five years; thus it is not surprising that only about one-third would recommend farming to a child or other relative.

Household Adjustments

Table 7.1 presents the proportion of the households who reported making household adjustments in the past five years (1984-1988) because of financial need. The 17 items are listed by rank for ease of presentation. The most frequently cited adjustments among farm families were postponing major household purchases (56 percent), using savings to meet living expenses (49 percent), and cutting back on charitable contributions (45 percent). Other frequently cited adjustments were changes in food shopping or eating habits (40 percent), changes in transportation patterns (40 percent), and spouses taking off-farm employment (38 percent). Nearly one-third of the households reported reducing utility use, taking off-farm work, saving less money for children's education, and postponing medical or dental care. About one-fourth reported using more credit, selling possessions, cashing in insurance, or reducing insurance coverage.

Findings: Does Financial Stress Make a Difference?

Perceptions of Well-Being and Debt-Asset Ratios

Table 7.2 examines the relationship between the four dimensions of familial well-being and the level of economic hardship as measured by farm debt-asset ratios. Respondents' perceptions of their quality of life were related significantly to the financial status of their farms. Farm operators and spouses with debt-asset ratios greater than 70 percent were much more likely to believe their quality of life had worsened over the past five years than those with debt-asset ratios of less than 10 percent. Thirty-one percent of the operators and 32 percent of the spouses on farms with debt-asset ratios greater than 70 percent, reported that their quality of life had become either somewhat worse or much worse over the past five years, compared with 13 percent of the respondents with debt-asset ratios of less than 11 percent. The relationship between debt-asset ratios and perceptions of declining quality of life were not related to the respondent's gender.

Nearly four of ten operators (41 percent) with debt-asset ratios greater than 70 percent reported that their satisfaction with farming had declined, compared with only 20 percent of those with debt-asset ratios of less than 11 percent. Likewise, spouses showed a strong positive relationship between satisfaction with farming and debt-asset ratios. Nearly one-half of the spouses (48 percent) with high levels of financial hardship reported that their satisfaction with farming had declined, compared with 23 percent of those with debt-asset ratios of less than 11 percent.

Respondents' feelings of regret if they had to move were unrelated to the financial condition of their farms. Fewer than 8 percent of both the operators and the spouses reported that they would not be sorry if they

TABLE 7.1 Adjustments Reported as Made Because of Financial Need in Past Five Years

	Percent Reporting
Postponed major household purchases	56
Used savings to meet living expenses	49
Cut back on charitable contributions	45
Changed food shopping or eating habits to save money	40
Changed transportation patterns to save money	40
Spouse took off-farm employment	38
Reduced household utility use, such as electricity and telephone	36
Changed transportation patterns to save money	35
Decreased the money saved for children's education	32
Postponed medical or dental care to save money	31
Sold possessions or cashed in insurance	25
Purchased more items on credit	25
Canceled or reduced medical insurance coverage	19
Fell behind in paying bills	19
Borrowed money from relatives or friends	16
Let life insurance lapse	15
Children postponed education	7

TABLE 7.2 Relationship Between Debt-Asset Ratio and Quality of Life Measures

	Debt-Asset Ratio			
	< 10	11-40	41-70	71 +
	------percent------			
Perceptions of quality of life worsening				
Operator	13	16	20	31
Spouse	13	16	20	32
Satisfaction with farming worsening				
Operator	20	28	31	41
Spouse	23	30	30	48
Not sorry about moving				
Operator	4	3	4	41
Spouse	7	4	6	48
Would not recommend farming				
Operator	45	43	44	49
Spouse	30	34	37	40

Includes respondents who indicated "become somewhat worse" or "much worse."

had to move. Because such a small proportion of the respondents stated that they would not be sorry, debt-asset ratios were not associated with these feelings.

Recommending farming to a child or other relative also was not related to the financial condition of respondents' farms. Regardless of the debt-asset ratio, 43 to 49 percent of the operators said they would not recommend farming to a child or other relative. Between 30 and 40 percent of the spouses said they would not do so. Spouses from farms with higher levels of economic hardship were more likely to state that they would not recommend farming than were spouses from farms in strong financial condition.

It appears that debt-asset ratios are associated strongly with perceptions that the quality of life is worsening and with declines in satisfaction with farming. Debt-asset ratios, however, were not accurate predictors of regret about leaving the community or recommending farming to a child or other relative.

Perceptions of Well-Being and Net Family Income

Each of the four measures of familial well-being was related to net family income (Table 7.3). The relationships were independent of the respondent's gender. Among both operators and spouses who suffered a loss in net family income in 1988, one-third reported that their family's quality of life had declined over the past five years, in contrast to fewer than 10 percent of those whose net family income exceeded $50,000.

Forty-two percent of the respondents who reported an income loss for 1988 also reported that their satisfaction with farming had declined, compared with about one-fifth who reported incomes greater than $50,000.

Income levels were related only marginally to respondents' feelings of not being sorry if they had to move. Ten percent of those reporting income loss said they would not be sorry if they had to move, compared with 4 percent of those with incomes greater than $50,000.

Approximately one-half of the respondents reporting income loss for 1988 stated that they would not recommend farming to a child or other relative, compared with 40 percent of the operators and 32 percent of the spouses whose family incomes exceeded $50,000. Even respondents with high incomes displayed a substantial reluctance to recommend farming to a child or other relative.

Household Adjustments and Debt-Asset Ratios

The 17 household adjustments revealed a consistent pattern: the

TABLE 7.3 Relationship Between Income and Quality of Life Measures

	Loss Was Realized	<$10,000	$10,000-$29,999	$30,000-$49,999	>$50,000
			----percent----		
Perceptions of quality of life worsening					
Operator	34	25	18	11	9
Spouse	33	24	18	10	7
Satisfaction with farming worsening					
Operator	42	35	30	21	17
Spouse	42	35	29	23	21
Not sorry about moving					
Operator	10	7	6	4	4
Spouse	10	8	7	7	4
Would not recommend farming					
Operator	55	55	48	42	40
Spouse	49	38	39	37	32

greater the economic hardship, as measured by debt-asset ratio, the higher the proportion of respondents stating they had made the adjustment (Table 7.4). For each measure, economic hardship was associated with making these adjustments. Eighty-two percent of those with debt-asset ratios greater than 70 percent reported postponing household purchases, compared with 53 percent of those with debt-asset ratios of less than 11 percent. Using savings to meet living expenses, reducing charitable contributions, and changing food shopping and eating habits were the most frequently mentioned adjustments.

Among respondents in strong financial condition (debt-asset ratios of less than 11 percent), a significant proportion reported making these adjustments. An insidious consequence of economic hardship in farming is that it permeated all farms, regardless of their financial condition. Even farm families in strong financial condition, defined as having debt-asset ratios of less than 11 percent, reduced expenditures, postponed purchases, and made other types of adjustments.

Relationship Between Perceptions of Well-Being and Household Adjustments

To examine the relationship between the four indicators of familial well-being and household adjustments, we summed the 17 individual adjustments into an index so that interval-level statistics could be

TABLE 7.4 Relationship Between Debt-Asset Ratio and Household Adjustments

	Total Sample	Debt-Asset Ratio			
		< 10	11-40	41-70	71+
		----------percent making adjustment----------			
Postponed major household purchases	56	53	62	71	82*
Used savings to meet living expenses	49	39	51	59	71*
Cut back on charitable contributions	45	41	45	54	64*
Changed food shopping or eating habits to save money	40	36	38	50	60*
Changed transportation patterns to save money	40	40	39	49	62*
Spouse took off-farm employment	38	32	40	53	62*
Reduced household utility use, such as electricity and phone	36	41	36	39	49*
Operator took off-farm employment	35	32	35	44	52*
Decreased money saved for children's education	32	26	34	44	57*
Postponed medical or dental care to save money	31	29	32	39	52*
Sold possessions or cashed in insurance	25	21	26	41	54*
Purchased more items on credit	25	18	28	40	46*
Canceled or reduced medical insurance coverage	19	17	20	25	31*
Fell behind in paying bills	19	10	19	31	53*
Borrowed money from relatives or friends	16	10	19	26	33*
Let life insurance lapse	15	13	14	21	30*
Children postponed education	7	8	7	7	14*

*p ≤ .05

computed. Perceptions of familial well-being were related statistically to the mean number of household adjustments (Table 7.5). For both operators and spouses, mean adjustment scores were higher among those who perceived their family's quality of life as worsening, reported diminished satisfaction with farming, would not be sorry if they had to move, and would not recommend farming to a child or relative.

TABLE 7.5 Relationship Between Household Adjustments and Perceived Quality of Life and Satisfaction

	Operator Mean Score	Spouse Mean Score
Perceived Quality of Life		
Much better	2.8	3.3
Somewhat better	3.7	4.2
Remained the same	4.8	4.8
Somewhat worse	8.0	8.3
Much worse	9.9	10.3
F-ratio	188.15**	148.26**
Satisfaction with Farming		
Much better	3.0	3.5
Somewhat better	3.9	4.0
Remained the same	4.2	4.5
Somewhat worse	6.7	6.4
Much worse	9.0	9.1
F-ratio	149.97**	104.41**
If you had to move, how sorry?		
Very sorry	4.9	5.1
Somewhat sorry	5.4	5.3
A little sorry	6.0	6.1
Not at all sorry	7.0	6.7
F-ratio	22.61**	11.72**
Would you recommend farming to child or relative?		
Yes	3.5	3.8
No	5.8	5.9
Not sure	5.1	5.4
F-ratio	126.21**	76.11**

** $p \leq .01$

Operators who perceived their family's quality of life as becoming much worse over the past five years had an average adjustment score of 9.9, compared with 2.8 for those who reported that their family's quality of life had become much better. Similarly, among spouses a strong relationship existed between perceptions of quality of life and the number of adjustments.

Operators who reported that their satisfaction with farming had become much better over the past five years had an average adjustment score of 3.0, compared with 9.0 for those who stated that their satisfaction had become much worse. The same pattern was observed among the spouses.

Those who reported that they would be "very sorry" if they had to move had lower household adjustment scores than those who said they would be "not at all sorry." Among both the operators and the spouses, feelings of not being sorry if they had to move were related positively to the number of adjustments they had made. Similarly, respondents who said they would not recommend farming to a relative or child had mean adjustment scores of 5.8, compared with 3.5 for those who stated they would recommend farming.

Who Was Most Likely to Make Household Adjustments?

The most accurate predictors of household adjustments were respondents' age, net family income, and debt-asset ratio (Table 7.6). Younger respondents with low incomes and high debt-asset ratios reported making the most adjustments. Operators under age 35 had an average adjustment score of 5.4, compared with 3.9 for those 55 or older. On farms where a loss of net family income was reported, the average adjustment score was 8.0, in contrast to only 3.4 for those with incomes greater than $50,000. Operators with debt-asset ratios of less than 11 percent had a mean adjustment score of 4.6, compared with 5.4 for those with debt-asset ratios between 11 and 40 percent, 6.9 for those in the 41-70 percent category, and 8.7 for those with debt-asset ratios greater than 70 percent. Respondents' educational levels, acres operated, and gross farm sales were not important determinants of household adjustments.

Conclusions and Implications

One of the difficulties in responding to the farm crisis was that it occurred unevenly. Some farm families were affected in 1980 or 1981; others were immune for three to five years after it began. The farm crisis evolved from a crisis into a chronic condition, and soon engulfed an ever-widening number of families. On some farms, economic hardship may have been precipitated by a particular situation or event such as a crop failure, unexpected bills due to machinery breakdown, uninsured losses, or illness. On other farms, perhaps no single event was responsible for financial hardship, but some families may have found themselves sliding into insolvency over a period of years as bills piled up, debts accumulated, and farm prices did not recover. Other families reported that they thought they were doing well until they received a notice from their lender, or until their lender refused to extend a mortgage or lend additional money. Some families were less vulnerable

TABLE 7.6 Relationship Between Socioeconomic Characteristics and Household Adjustments, Zero-Order Correlations (data uncategorized with exception of gross farm sales)

Personal Characteristics	Household Adjustment Mean Score	
Operator's Age		
≤ 35	5.4	F-ratio = 87.79**
36 to 55	5.8	
> 55	3.9	
Operator's Education		
< 12 years	4.4	F-ratio = 87.62**
12 years	5.0	
13 to 15 years	5.6	
≥ 16 years	4.8	
Spouse Age		
≤ 35	5.9	F-ratio = 94.45**
36 to 55	5.6	
> 55	3.7	
Spouse's Education		
< 12 years	4.6	F-ratio = 2.63*
12 years	5.1	
13 to 15 years	5.1	
≥ 16 years	4.7	
Net Family Income (1988)		
Loss was realized	8.0	F-ratio = 70.44**
< $10,000	6.2	
$10,000 - $29,999	5.2	
$30,000 - $49,999	4.1	
≥ $50,000	3.4	
Farm Characteristics		
Acres Operated		
≤ 160 acres	5.2	F-ratio = 1.72
161 to 300 acres	5.7	
301 to 640 acres	5.1	
> 640 acres	4.9	
Gross Farm Sales (categorized)		
< $10,000	5.2	F-ratio = 7.06**
$10,000 to $39,999	5.2	
$40,000 to $99,999	5.1	
$100,000 to $249,999	4.6	
≥ $250,000	4.0	
Debt-Asset Ratio		
≤ 10	4.6	F-ratio = 80.06**
11 to 40	5.4	
41 to 70	6.9	
≥ 71	8.7	

* Significant at .05
** Significant at .01

or were affected less than others, but the impact of the farm crisis was uneven within communities and among farm families. These differences in effects have created three somewhat distinct subgroups of farmers: those in strong financial positions, those in very weak positions, and those who are "getting by."

In the early 1980s, these groups were nearly equal in size, with approximately one-third in each category (Harl 1987). As financial conditions improved during the late 1980s, these categories remained, although a smaller proportion now belongs to the high-distress category. The high-distress category is smaller because many families simply could not hang on and quit farming as a result of economic hardship, or because improved economic conditions may have enabled some families to pay off loans, restructure their debts, and move onto sounder financial footing. The adage "a rising tide lifts all boats" is appropriate in describing the improvement. The gap appears to be widening, however, between those in strong financial condition and those at high levels of economic distress. In these data, about four respondents in ten reported that their family's quality of life had improved over the past five years; on the other hand, nearly two in five reported declines in their quality of life. This distinction is evident also in the respondents' level of satisfaction with farming: about one-fourth reported improvement in job satisfaction and about one-fourth reported a decline.

This widening gap among farm families is striking when financial data are used to classify respondents. Three times the proportion of those with debt-asset ratios greater than 70 percent were likely to report a worsening in their quality of life, and were twice as likely to report declines in job satisfaction. The same pattern emerged when we used net family incomes to classify respondents.

As expected, household adjustments were related positively to levels of economic hardship. In each of the 17 adjustments, higher debt-asset ratios were associated with making these adjustments. Among respondents with ratios greater than 70 percent, the average adjustment score was 8.7, compared with 4.6 for those with ratios of less than 10 percent. One limitation in the presentation of the household adjustment data is that all of the items were weighted equally. Some of the adjustments, however, are more serious than others, although the proportion of respondents who made them may be smaller. For example, a "major household purchase" includes a wide variety of goods. Postponing purchasing a new sofa and getting by with the old one probably would be less disruptive than postponing buying a new refrigerator to replace a broken one. It is not possible, however, to disaggregate the data for this type of analysis. In addition, although 56 percent of the entire sample postponed major household purchases, this type of adjustment may be less damaging to familial well-being than postponing medical or dental care, which was reported by 31 percent of the sample. When one considers the long-term health implications of postponing health care, these hidden costs could well exceed the hoped for savings. Future research should examine how important the adjustments are to respondents. Using savings to meet living expenses may not be traumatic for some families, but may be very painful for those who used savings that were intended to finance their children's education. These qualitative distinctions between household adjustments should be incorporated into future research.

Household adjustments were associated with the respondents' financial status; these findings support the research reported by Bultena, Lasley, and Geller (1986). Unexpected, however, was the relatively high incidence of these adjustments, even among those in strong financial condition. Borrowing money from relatives, canceling medical insurance, and postponing medical care and children's education were coping strategies employed by a significant proportion of the respondents regardless of financial status. Perhaps the data are tapping into the informal economy that exists among farm families. More research attention should be given to the extent to which farm families borrow money from relatives, hold multiple jobs, start their own businesses, and employ other strategies to remain on their farms. Again, the limitations of the data set do not allow more than speculation about these dimensions.

Several plausible explanations may account for the relative high incidence of adjustments among farm families in strong financial condition. Possibly they feared that they, too, might become victims of the crisis, and thus made adjustments as preventive measures. For some, these adjustments may have reflected a strategy of "hunkering down until the storm passes." Others may have made these adjustments not because of personal economic need, but because of empathy or concern about what they were learning in the media or seeing in their own neighborhood. One important distinction between those who were forced to make these adjustments and those who made them voluntarily is the matter of choice. Choosing to postpone a purchase, to take an off-farm job, or to change spending habits is considerably different from being forced to take these steps because of economic need or lack of any other options. Although the questionnaire asked respondents to indicate those adjustments they had made because of economic need, the data show that persons in strong financial positions were making them for other reasons.

The data support the prediction that assessments of quality of life and job satisfaction are related negatively to making household adjustments. For example, those who believed that their family's quality of life had declined in the past five years had an average adjustment score of 10, compared with 2.8 for those who believed it had improved greatly. Adjustment scores showed a statistically significant difference across the four indicators of familial well-being.

The widening gap in financial status between those at the top and those at the bottom is reflected in their assessments of their family's well-being. Assessments of quality of life, opportunities, and choices are affected by financial conditions. Future research should focus on how the experiences of financial hardship shape lifelong views about farm life. Perhaps the large proportion of the respondents who were not willing to recommend farming to a child or other relative reflects a growing pessimism about the prospects for farm life. If the financial hardship continues, how will it shape the structure of agriculture and the social fabric in farming communities? What will happen to the family farm structure of agriculture when too few young people are attracted to the industry to replace those who retire? How will the failure of a generational turnover in farming affect rural communities as the number of farms declines further and the size of farms increases?

The data show that the great majority of respondents, independent of gender, would be sorry if they had to leave their communities. Even those in dire economic conditions expressed a significant level of community attachment. Future research should explore whether these feelings of attachment are shared by the respondents' children. Although this point is only speculative, perhaps commitment to farming and attachment to the community are not being instilled in the next generation, in view of the proportion of respondents who say they would not recommend farming. Future research should focus on the values being transmitted to farm youths. Are they receiving the message that farming communities are a good place to live but a poor place to make a living? What message are young people receiving from their parents' inconsistent attitudes and behaviors? Future research should explore how the decade of the farm crisis influenced farm youths' occupational and career choices. The lessons of the 1980s will live long in the memories of the young people who witnessed the household adjustments made by their parents throughout this turbulent period.

References

Banks, M. H. and P. R. Jackson. 1982. "Minor Psychiatric Disorders in Young People: Cross-Sectional and Longitudinal Evidence." *Psychological Medicine* (12): 789-798.

Barrett, J. 1987. *Mending a Broken Heartland*. Alexandria, VA: Capitol Publications.

Brenner, M. H. 1973. *Mental Illness and the Economy*. Cambridge, MA: Harvard University Press.

Bultena, G., P. Lasley, and J. Geller. 1986. "The Farm Crisis: Patterns and Impacts of Financial Distress among Iowa Farm Families." *Rural Sociology* 51(4): 436-448.

Cobb, S. and S. V. Kasl. 1977. "Termination: The Consequences of Job Loss." Behavioral and Motivational Factors Research Report 76-1261. Washington, DC: National Institute for Occupational Safety and Health.

Dooley, D. and R. Catalano. 1980. "Economic Change as a Cause of Behavior Disorders." *Psychological Bulletin* 87: 450-468.

Geller, J., G. Bultena, and P. Lasley. 1988. "Stress on the Farm: A Test of the Life-Events Perspective among Iowa Farm Operators." *Journal of Rural Health* (2): 43-57.

Harl, Neil. 1987. "The Financial Crisis in the United States." Pp. 112-129 in *Is There a Moral Obligation to Save the Family Farm?*, edited by Gary Comstock. Ames: Iowa State University Press.

Heffernan, W. D. and J. B. Heffernan. 1986. "Sociological Needs of Farmers Facing Severe Economic Problems." Pp. 90-102 in *Increasing Understanding of Public Problems and Policies*. Oak Brook, IL: Farm Foundation.

Prairiefire. 1987. *Renew the Spirit of My People: A Handbook for Ministry in Time of Rural Crisis*. Des Moines, IA: Prairiefire Rural Action, Inc.

Rosenblatt, Paul C. and Roxanne M. Anderson. 1981. "Interaction in Farm Families: Tension and Stress." Pp. 147-166 in *The Family in Rural Society*, edited by R. T. Coward and W. Smith, Jr. Boulder: Westview.

8

Perceiving Hardship and Managing Life

Katherine Meyer

The restructuring of the economy brought changes in agriculture and in the American farmer's overall well-being. The productive value of the family farm suffered a relatively abrupt reduction, which reduced its capacity to consume as well. The media documented signs of this change before social scientists began to study seriously what was happening in farm life. The popular press presented several documentaries and the film industry depicted crisis on the farm in films such as *The River* and *Country*. All of these documentaries and realistic films highlighted features of farm life in the 1980s which resonated with America's general understanding of the plight of its farmers. Abandoned schools, deserted farmsteads, closed churches, and boarded-up businesses were common sights for tourists passing through the Midwest. Personal and individual struggles against mortgage foreclosures were reported on the evening news. So, too, were protests of farmers in front of local banks and the halls of the United States Congress. The cameras showed the weary faces of farm men and women stunned by what was happening to a way of life that most of them had lived for generations.

Events for America's farmers were not unlike events in other sectors of the population. Plant closings hit towns in the Northeast and the North Central regions as industries moved to the South or the Southwest, or other countries. The media portrayed the process, the personal struggles, the protests, and the outcomes. Social scientists studied plant closings with attention to the consequences. Social psychological effects, as well as economic and community outcomes, were explored (Hass 1985; Kasl and Cobb 1979; Kinichi 1985; Mick 1975; and Perruci and Targ 1988). Workers had become more conscious of themselves as a social class with distinct interests. They reported stress, economic distress, demoralization, and depression.

Nuclear and chemical environmental disasters also captured national attention during the 1980s. Researchers focused on hazardous toxic waste and toxic exposure as it impacted residential communities (Bachrach and Zautra 1985; Cook 1983; Edelstein 1988, 1989; Edelstein and Wandersman 1987). Their findings demonstrated that there was consistency in how communities dealt with disaster. The population mobilized or remobilized to deal with these new, life-threatening problems. Social movements showed signs of formation and growth. People formed and joined groups. New organizations emerged; existing organizations became more activist. Political involvement, awareness, and activity were evident. Leaders were identified; goals and ideas were articulated; the media were employed; local, state, and sometimes national resources were put to use. Along with the picture of an energetic, goal-oriented, and mobilized population, the research documented the stress, the symptoms of mental health problems, and the toll exacted from human beings as they organized and dealt with an environmental disaster or its threat.

Studies of plant closings and of contaminated communities echoed themes prevalent in research on community responses to natural disasters such as earthquakes, floods, and hurricanes. During the late 1960s and the 1970s, disaster research increased exponentially. In an article summarizing research findings to date, Quarantelli and Dynes (1977) identified strengths and weaknesses of disaster research and gaps in information which still needed to be filled. In doing so, they articulated trends that now seem familiar when we consider dramatic changes. The research on disasters had shifted its emphasis from social psychological outcomes to a context of social organization and reorganization. A systems perspective was applied to empirical findings. Links between group behavior in disasters and behavior in complex organizations erased the long-held notion that conventional behavior differed ontologically from collective behavior. The period preceding disasters was studied as well as the disaster itself and the aftermath. Research now focused on long-range outcomes in affected communities and in their populations.

It would be incorrect and simplistic to suggest that fully integrated sets of findings are available to researchers interested in the process and the consequences of social changes. Each area of research—plant closings, environmental contamination, and natural disasters—documented its own inconsistent findings, its struggles to make theoretical sense of descriptive data, its empirical gaps, and its points of debate. Research on populations at risk has recurrent emphases and themes but this does not reduce the problems for those of us who are interested in the impact of rural restructuring on the farm population of the Midwest. It implies, however, that we are not in uncharted territory; we can be guided by the insights of other observers of social change. It suggests further that when we learn about the farm population, everyone learns more about at-risk populations in general. The hope guiding the case studies is that the sum total of these studies will provide cumulative evidence, in this instance, of the human effects of structural change.

Social Psychological Effects of Structural Reorganization

Each of the bodies of research cited above examined social psychological effects as well as community or structural effects of major disorganizing events. Research on job loss and unemployment did so as well (Bowman 1988; Snyder and Nowak 1984). In other studies, rapid growth was the change event (e.g., Freudenburg 1982). All of these studies explored consequences such as depression, anxiety, stress, or other indicators of personal distress.

While research on disasters, economic disruption, and environmental threat was developing, social scientists accumulated a considerable body of information that addressed indicators of emotion. A vast literature on stress emerged (Kobasa and Puccetti 1983). Depression was discussed as both a biomedical entity (Baldessarini 1983; Gold, Goodwin, and Chrousas 1988) and a social fact (Mirowsky and Ross 1986; Thoits 1981). Literature also was devoted to variables that could reduce stress, depression, or both, such as coping techniques (Billings and Moos 1981; Lazarus and Folkman 1984; Miller and Green 1985; Pearlin and Aneshensel 1986; Stone and Neale 1984; Wheaton 1982, 1985). Similarly, social support was the subject of much research (Cohen, Sherrod, and Clark 1986; Eckenrode 1983; Jung 1984; Kessler and McLeod 1985; Wethington and Kessler 1986). Control was analyzed alone and in combination with support (Krause and Stryker 1984; Lefcourt, Martin, and Saleh 1984; Ross and Mirowsky 1989). The literature on emotion often was linked to changes in individuals' lives (Cohen, McGowan, et al. 1984; DeLongis et al. 1982; Hammen et al. 1981). By implication, either the social structure was static or the change in the structure was not dramatic. The literature seldom has been linked to change events that affect a whole group. An important theoretical breakthrough will have occurred when change experienced by a group is analyzed with the fullness of insight acquired from research on stress, depression, and their mediators.

Some of the insight provided by the research on emotions has come from careful attention to gender differences in stress and depression (Al-Issa 1982; Amenson and Lewinsohn 1981; Barnett, Brener, and Baruch 1987; Belle 1982; Briscoe 1982; Thoits 1985; Warr and Parry 1985; Williams 1985). In part, the differences in stress and depression experienced by men and women were explained by differences in the models of support and/or control and coping which are available to, or used by, men and women in American culture. As studies of change experienced by a group explore social psychological outcomes, they must include gender differences.

Rural Transformation and Mental Health

Some of the research on farm populations has examined the restructuring of the rural sector and the well-being of farmers. A number of important studies have addressed rural transformation and mental health (Belyea and Lobao 1990; Duncan, Volk, and Lewis 1988; Keating 1987; Keating, Munro, and Doherty 1988; Rosenblat and Keller 1983; Staten 1987). Some of them attended explicitly to the theoretical literature on stress and depression (Davis-Brown and Salamon 1988). Others paid special attention to gender or the well-being of farm women (Haney and Knowles 1988; Lobao and Meyer 1990; Tigges and Rosenfeld 1987; Whatmore 1991). All of these studies have contributed to resolving what remains a knotty problem. How do we understand gender differences in stress and depression in populations that simultaneously are experiencing, in varying degrees, a change such as rural restructuring? Much of our understanding is limited to what we know from case studies of individual communities or states. Little is discussed in the context of what is known about emotion in non-rural populations.

One goal of the North Central Regional Farm Survey was to document stress, coping, and depression in men, women, or both as the U.S. rural sector was transformed during the 1980s. We deliberately built into the survey design a regional rather than a case study sample. We included questions important not only to students of rural life but to those who study mental health and those who study various populations at risk because of rapid change.

Stress, Coping, and Depression
Among Midwest Farmers

Because of multiple research interests and limitations on questionnaire space, the study focused on particular elements of Midwestern farmers' mental health during the economic and social turmoil of the 1980s. Both operators and their spouses were asked about their perceptions of economic hardship over the past five years in questions such as whether their own family's finances had improved, and whether their financial situation was better than that of other farmers in their area. Men and women also were asked whether their level of stress had increased, how much they were concerned with stress, and how much stress they felt day to day.

Additional questions were asked only of women for several reasons. Women often are viewed as buffers for economic effects on family units (Lobao and Meyer 1991). In the division of labor, they also are typically involved with consumption for their families. Thus we wanted to assess their perceptions of effects on the family unit in order to learn something about the consequences of change and turmoil. Also, one of the research goals was to increase our understanding of rural mental health by testing some preliminary hypotheses about the connections between personal well-being and economic change that affects a whole group. Given the limitations of time and space, we decided to test a few questions only on the women. As a research team, the principal investigators of this study had affirmed that the family farm would serve as the research unit. Both operators and their spouses, they stated, could report equally well when information about the farming operation was needed. On some issues, it would be sufficient to ask either spouse about what was happening. Both partners would be consulted, however, when we wanted to compare experiences that were unique to either the operator or the spouse.

As the mediators of economic stress or distress, women were asked about the pressures on the farm family unit which they experienced. They reported how often problems occurred in balancing work with family duties, in making child care arrangements, in indebtedness and debt servicing, in having and keeping farm help, in adjusting to new government policies, in lacking control over weather and commodity prices, in finding support from spouses in farm or family duties, and in conflict with spouses and children. We also asked women whether their families experienced hardship as consumers of food, clothing, and medical care—basic household needs. We included items from the Center for Epidemiological Studies Depression Scale to assess the extent of women's depression on a generally accepted measure of depression. Women were asked whether they were without hope, restless, unhappy, lonely, unusually bothered by things, unable to concentrate, depressed, sad, finding everything to be an effort, or unable to enjoy life.

Coping strategies often are viewed as mediators between change events, on one hand, and depression and stress, on the other. We asked women how often they used each of 18 mechanisms to deal with serious farm problems. The list of items presented to the Midwestern women was derived from a list of 37 items used by Lobao and Meyer (1990), in a study of men and women farmers in Ohio. In that study, the authors factor analyzed items that they constructed or drew from scales published by Lazarus (1981) and Belyea and Lobao (1990). For the North Central study, Meyer and Lobao selected those items which had been most representative of different styles of coping among Ohio farm women.

Measures of Stress and of Economic and Social Hardship

Principal components factor analysis with varimax rotations proved a conservative and effective way of constructing all scales or multi-item indicators of stress, depression, coping, and economic and social hardship. Co-variation among items measuring stress was high for both men and women; analysis of this co-variation explained 86 percent of the variation in the three individual items intended to estimate stress. In examining frequency distributions for both men and women, we found that about 58 percent of the men reported increased stress over a 5-year period, as did about 50 percent of the women. Forty-two percent of the women perceived their stress to be about the same over time; 31 percent of the men did so. Very small percentages of men (11 percent) and of women (8 percent) reported any decline in stress. According to multiple-item indicator of stress, Midwestern farmers felt that stress had worsened or remained stable for them. A slightly greater percentage of men than of women reported increased stress.

In an analysis of how men and women viewed their economic situations or family finances, 43 percent of the men and 41 percent of the women believed their family finances had improved over the past five years. Twenty-eight percent of the men and 26 percent of the women thought their finances were about the same. Similar consistency in evaluating their economic condition was evident in men's and women's comparison of their finances to those of other farmers in their area: 33 percent of the men and 29 percent of the women thought their financial situation had improved more than other farmers' situations. Fifty-two percent of the men and 51 percent of the women saw their financial situations as remaining the same while that of other farmers fluctuated. Our sample of Midwestern farmers, however, was drawn from those who had remained in farming and had not left during the 1980s. That fact affected their perception of their own economic well-being. The consistency between the operator and the spouse samples also is important: the two were virtually identical in their reporting of family finances over a 5-year period, and in comparing their situation to that of other farmers in their area. That consistency was convincing. The reports of family finances and comparative finances also were reliable across gender divisions. In addition, we found support for an assumption on which we based parts of our methodology in this research: operators and spouses reported similar information about the farm operation when data were aggregated. That finding, of course does not imply that each operator and each spouse individually agreed in the same way.

Information about pressures, coping, depression, and consumer hardship was available only on farm women. Difficulty in affording food, clothing, and medical care over the past twelve months worked as a set of variables. Association among these three indicators was high; factor analysis of their co-variation produced a factor that we called "consumer hardship," which accounted for 84 percent of the variation in the items taken as a set. Consumer hardship, however, was a highly skewed variable: 61 percent of the women reported that they never experienced hardship. Another 18 percent said they did not experience it often. Only about 8 percent of the total experienced it fairly often or very often. These findings reminded us that the sample consisted of farm survivors. Most of the men and women reported improved or stable financial conditions. Most of the women reported that household consumption, typically managed by women, had not suffered very much.

We examined the daily pressures that women faced (Table 8.1). Factor analysis showed that the pressures could be grouped into three sets: those coming from family duties and relationships, those connected with the farm operation, and those coming from external economic sources. The first set included conflict with spouse or children, degree of support from spouse, and striking the balance between work and family duties. (Respondents had been asked about difficulty with child care arrangements, but because of the age range of the women in the sample, this was not a relevant issue for 71 percent. Therefore, we dropped this item from further analysis.) The second set of pressures encompassed adjusting to new government policies, problems with hired labor and fluctuations in weather and commodity prices. A third pressure came from debt or indebtedness.

Some women (32 percent) reported that they almost never

TABLE 8.1 Sources of Daily Pressures: Rotated Factor Pattern

Variables	Family Duties and Relationships	Farm Operation	External Economy
Balancing work and family duties	.5868	.2199	.2919
Conflict with spouse	.8300	.0968	.1184
Conflict with children	.7459	.0074	-.1374
Adjusting to new government policies	.1442	.7107	.0242
No farm help/loss of help when needed	.0905	.7283	-.1946
Lacking control – weather and commodity prices	.1532	.6861	.2760
Insufficient spousal support in farm/family	.5848	.3497	-.0388
Indebtedness/Debt	.0292	-.0109	.9275

experienced pressures from family duties and relationships. Another 50 percent said such pressures were occasional. Only about 10 percent said they occurred daily. (Some women reported that family pressures did not apply to them.) Closer investigation of items included in the scale confirmed the impression that more than 80 percent never or rarely suffered family-related pressures. The only item for which a sizable proportion of women (23 percent) reported pressure was the daily juggling act to balance household and farm roles.

Neither did farm operation pressures seem overwhelming: 24 percent of the women almost never felt such pressures, and an additional 48 percent felt them only occasionally. Only 9 percent reported daily farm pressures. The only single item in the scale for which a sizable number of women reported daily pressure was the lack of control over weather and commodity prices (26 percent). If we interpret the distribution of the uncertainty about weather and commodity prices in conjunction with the item on juggling home and farm duties, it seems that women experience more severe pressure from unpredictability and uncertainty than from particular divisions into family-related and operation-related farm labor.

To summarize, the farmers generally viewed their financial situation as stable or improved. Farm women reported that the ability to consume also was relatively steady. The women did not seem to be buffeted by pressures emanating from family farm life. Both men and women, however, reported that stress had remained constant or had increased. Women suffered pressure from unpredictable aspects of family life, or the farm operation. Although it would be incorrect to portray most Midwestern farmers as unduly buffeted and stressed and as keenly aware of rapid declines in their fortunes and spending ability, it would be equally incorrect to view those who have remained in farming as tranquil and worry-free.

Measures of Depression and Coping

We measured depression with a shortened version of the CES-D scale, following the example of Ritter et al. (1990), Ross and Mirowsky (1989), and earlier work by Ross, Mirowsky, and various collaborators. Our shortened version included items that worked well together in a scale of Ohio farm men and women, as well as a few items that tapped women's depression more accurately than men's. Women were asked "How many days during the past week have you: 1) felt without hope, 2) had restless sleep, 3) felt unhappy, 4) felt lonely, 5) been bothered by things that usually aren't usually bothersome, 6) had trouble concentrating, 7) felt depressed, 8) felt that everything was an effort, 9) felt sad, and 10) failed to enjoy life?" We summed the responses to produce an index of depression. Factor analysis of the index supported a single-factor solution with internal consistency of .72.

Only a very small percentage of the women (fewer than 3 percent) reported consistent depression. Analysis of individual items supported the interpretation that the farm women sampled had a basically positive attitude. Depression, when reported, was only an occasional state. Forty-four percent of the sample were rarely depressed, and that percentage would have been higher if the items on restless sleep and hopelessness had not been included. About half of the women reported that they always enjoyed life and were rarely depressed, bothered, or lonely. If we treat responses of "sometimes" or "occasionally" as a unit, 53 percent of the women reported some depression, but 97 percent of the women, altogether, were depressed only rarely or once in a while.

These findings are understandable when we recall the characteristics of the women sampled here. Their farm households were surviving; they were mostly in their middle years; all were spouses of farm operators. In addition, as reported above, most did not have young children in the home. In the Illinois Survey of Well-Being, Mirowsky and Ross (1986) found that economic well-being reduced depression. They also found that depression is generally lower among middle-aged persons, partly because of economic well-being. Married people are less distressed than unmarried people. Again, economic well-being is a factor, and social support is suggested by marital status. Mirowsky and Ross also reiterated a well-documented research finding: the presence of children in the home does not improve the parents' psychological well-being. Although the Illinois Survey of Well-Being is summarized here, its findings are consistent with many studies on depression, which include age, economic standing, marital status, and children's age, the presence of children in the home, or both. The Midwestern farm women's depression scores were consistent with those we would expect in a general non-farm sample of women. Men and women usually are compared, and the women have higher depression scores, but data on depression in men were not available from the Midwestern study.

Women were asked how often they used the 17 strategies for coping when they were faced with serious farm problems such as drought and low prices. Factor analysis of the 18 items yielded 4 defined factors (Table 8.2). One factor implied support-seeking strategies for coping (Factor 4). It included talking to someone who could solve the problem, speaking with a counselor, and looking for support from friends and relatives. Although the variable "talking to a counselor, or other mental health professional" grouped definitively on the factor with the other support-seeking strategies, analysis of individual items showed that 92 percent of the women reported never having used counseling. Thus we dropped it from further analysis. Sixty percent of the women used other social support techniques. (Seventy-three percent sought support from family and relatives; 53 percent sought spiritual support; 53 percent talked to someone they thought would be helpful). The median category for support seeking of any type was "somewhat." The women did not use social support frequently and consistently as a coping strategy; it was only an occasional mechanism.

A second factor (Factor 3) involved a behavioral component, activity. Items reporting participation in church activities and becoming more involved in activities outside the farm led the pattern of loadings. Analysis of individual items showed that 86 percent of the women participated in church activities and 86 percent became more involved in outside activities.

Factor analysis provided a third factor, which we can interpret as a denial or withdrawal technique of coping (Factor 1). The items that grouped included using potentially addictive behaviors to feel better, refusing to think about the problem, keeping feelings to oneself, keeping problems secret from others, wishing the problems would go away, and not expecting much income from farming. Interassociation of the items was .82.

Analysis of individual items in the scale revealed that 71 percent of the women used these techniques, except for eating, drinking, smoking, using medication, and the like, to feel better. (Sixty-six percent of the women reported never using those potentially addictive

TABLE 8.2 Coping Strategies: Quartimax Rotated Factor Pattern

Variables	Denial/ Withdrawal	Mental Control	Activity	Support-Seeking
Participating church activities	-.0401	.0484	.7665	.1015
Involving more outside activities	-.0311	.0743	.7054	.0436
Noticing others with more difficulties	.0871	.4011	.4822	-.0094
Telling self farm success not important	.1435	.5544	.3436	-.0225
Reminding self for good things happened	.0140	.7775	.1779	.0249
Putting up a lot to stay farming	.1470	.7400	.1779	.0249
Pretending nothing is happening	.4152	.5564	-.0371	-.0251
Making up plans and follow	-.1039	.4843	.2011	.2452
Using addictive ways to feel better	.4755	-.0569	-.1283	.3248
Refusing to think about it	.6421	.1670	-.0457	.0316
Keeping problems secret	.7251	.1112	-.0060	-.0976
Seeking family/relative support	-.0446	.0674	.2808	.6420
Seeking spiritual support from clergyman	-.0139	.1264	.5892	.4298
Talking to counselor	.1446	-.0326	-.0489	.5923
Not expecting much from farming	.5662	-.0195	.1086	-.0082
Talking to someone who can help	-.1151	.1981	.1344	.6565
Wishing situation away	.6414	.0526	-.0289	.1637

behaviors at all.) Although 71 percent reported coping through the other forms of denial or withdrawal, only 11 percent used those techniques a great deal. The modal category of response was "somewhat"; 44 percent of the women gave that answer. In a comparison of the ways of coping with problems, women reported using denial or withdrawal more than seeking social support (71 percent and 60 percent—respectively) but less than constructing mental or social activity to deal with problems (71 percent denial, as compared to 86 percent structuring).

A fourth factor focused on mental control and positive comparisons with others (Factor 2). Respondents noticed people who had more difficulties, told themselves that there was good and bad about farming, and put up with problems while making a living. Internal consistency among the items was .78. Eighty-six percent of the respondents used each of these techniques when facing problems. Eighty-eight percent reminded themselves of the good and bad in farming, and 94 percent noticed people who were worse off. Eighty-three percent told themselves that farm success wasn't the only thing in life, and 76 percent put up with problems to make a living. The modal categories for each response were "quite a bit" and "somewhat." If we compare respondents' use of this comparisons/tolerance factor with the increased activity factor, it is clear that women farmers structured reality both mentally and socially in order to cope with problems.

Farm Women and Coping

Past research on non-farm samples which examined coping strategies focused either on conceptualizing coping or on the outcomes. Outcome research has explained how different types of strategies were more or less effective in reducing depression or stress (e.g., Pearlin and Aneshensel 1986). Measurement research has sought to identify types of strategies that people used (Folkman and Lazarus 1980). The Midwestern farm women in this study were generally not depressed; our findings addressed how they coped with problems, not what techniques mitigated stress or depression for them. (Of course, there may have existed some connection that we could have explored more fully if we had more women who were depressed frequently—perhaps those who had been forced to leave farming.)

Other studies identified various of types of coping. Very broadly, those types could be classified as direct action, emotion management, and interpretive reappraisal. Ungrouped, specific types included confrontive coping, distancing, self-control, seeking social support, accepting responsibility, escape-avoidance, planful problem solving, positive reappraisal (Folkman et al. 1986), direct action, optimistic comparison, selective inattention, and restricted expectations (Pearlin and Lieberman 1979). The North Central Regional Farm Survey included items representing virtually all of these types of coping. The items were chosen from the most accurate indicators of various coping strategies in research on Ohio farm men and women (Meyer and Lobao 1991).

Midwestern farm women employed four coping strategies: reality construction, denial/escape, increased activity, and support seeking. Reality construction combined items reflecting selective inattention, optimistic comparison, and restricted expectations. Women farmers oriented themselves to problems through and a mental definition of the situation. Their behavior reflected what ethnomethodologists call the social construction of reality. Both in their minds and in their behavior, the women shaped the world around them.

Women farmers also denied and escaped problems. Typologies or coping strategies from other research suggested that distancing and escape/avoidance were separate strategies. For farm women, however, staying clear of problems was a total coping process, not one which could be divided into subcategories or types of coping. They didn't want to think about the problems, talk about them, admit them, tell anyone about them, or confront the situation. They didn't avoid problems, or distance themselves as separate kinds of responses; they did both. Women farmers sought support. The techniques they used reflected those found in research on coping in other populations, except that few spoke with counselors or mental health professionals. They sought social support from those persons close to them, religious leaders, and those who could be helpful. Studies of depression have demonstrated that support and control can substitute for one another as effective mechanisms for reducing depression (Mirowsky and Ross 1989). Although that hypothesis was not directly testable with the data on Midwestern farm women, reality construction (a way of controlling life events) and support seeking were used by the farm women, most of whom were only occasionally or never depressed.

Finally, they became active. Activity is often cited as an important way to diminish depression, although it can be stressful. Getting out of the house and getting involved outside of the home focus attention away from one's problems and difficulties.

The Midwestern Farmer: Enduring or Prevailing

Analysis of the perceptions and social psychological dispositions of Midwestern farmers has been both intriguing and tantalizing. Many questions were answered only partially or not at all. Although the women were not overwhelmingly depressed or stressed, what about their sense of self-esteem, efficacy, or trust? Each of these is as much an indicator of the mental state as are depression and stress. What about men? Their stress was higher than women's; what about their depression? Did men and women use the same coping strategies in the same order, or do farm men and women cope differently from each other? Do men have different coping strategies from the non-farm population? Do depressed and stressed men and women have common profiles, or is each of these psychological states unique to the farmer's gender? All of the members of this sample were still farming. What about those who left farming? Were stress and depression strong factors in that decision? Did those individuals cope with problems differently from the enduring farmers? Did they perceive hardship in the same way? How did they view their economic situation relative to that of others? Each of these questions deserves future research. Focusing only on economic outcomes of rural restructuring or only on community effects misses a crucial dimension. The media, the American people, and others who observe rural America know that what happens in farming is not detached from what happens to the farmers' social and psychological well-being. When factories close, when disaster strikes, when communities become toxic or contaminated, the account of events includes, by definition, what the affected people perceived, experienced, and felt.

This study of Midwestern farmers, particularly women, provides a basis for developing a profile of those who endured the restructuring of the rural sector during the 1980s. The survivors viewed themselves as better off than others; the women felt they still could manage to function as consumers, still could afford what their families and households needed. Yet they did not play down the stress involved—even if they were survivors. Most of the women were not depressed too much, but almost all were sometimes depressed and a small percentage were very depressed. Were those who left farming better off or not? The women also demonstrated styles of coping different from those found in non-farm populations, but incorporating aspects of those styles. Demographic characteristics may explain part of the absence of severe depression: we might infer that most women were happy because they were older, married, financially solid, and basically finished with child rearing. We also expect, however, that enduring women have well-developed mechanisms of solving, resolving, and living with problems—even serious ones.

References

Al-Issa, I., ed. 1982. *Gender and Psychopathology*. New York: Academic Press.
Amenson, C. S. and P. M. Lewinsohn. 1981. "An Investigation Into the Observed Sex Difference in Prevalence of Unipolar Depression." *Journal of Abnormal Psychology* 90: 1-13.
Aneshensel, C. S. 1985. "The Natural History of Depressive Symptoms: Implications For Psychiatric Epidemiology." *Research in Community and Mental Health* (Vol. 5)., edited by J. R. Greenley. Greenwich CT: JAI Press.

Bachrach, Kenneth and Alex Zautra. 1985. "Coping with a Community Stressor: The Threat of a Hazardous Water Facility." *Journal of Health and Social Behavior* 26: 127-141.

Baldessarini, Ross J. 1983. *Biomedical Aspects of Depression and Its Treatment*. Washington, DC: American Psychiatric Press.

Barnett, R., L. Brenner, and G. Baruch, eds. 1987. *Gender and Stress*. New York: Free Press.

Belle, D., ed. 1982. *Lives in Stress: Women and Depression*. Beverly Hills, CA: Sage.

Belyea, Michael and Linda Lobao. 1990. "Psychosocial Consequences of Agricultural Transformation: The Farm Crisis and Depression." *Rural Sociology* 55(1): 58-75.

Berkowitz, Alan D. and H. Wesley Perkins. 1984. "Stress Among Farm Women: Work and Family As Interacting Systems." *Journal of Marriage and the Family* (February): 161-166.

Billings, A. G. and R. H. Moos. 1981. "The Role of Coping Responses and Social Resources in Attenuating the Stress of Life Events." *Journal of Behavioral Medicine* 4: 139-157.

Bowman, Phillip J. 1988. "Postindustrial Displacement and Family Role Strains: Challenges to the Black Family." Pp. 75-97 in *Families and Economic Distress: Coping Strategies and Social Policy*, edited by Patricia Voydanoff and Linda C. Majka. Newbury Park, CA: Sage.

Briscoe, M. 1982. "Sex Differences in Psychological Well-Being." *Psychological Medicine*, 12, Monograph Supplement 1.

Cohen, S., J. McGowan, S. Fooskas, and S. Rose. 1984. "Positive Life Events and Social Support and the Relationship Between Life Stress and Psychological Disorder." *American Journal of Community Psychology* 12: 567-587.

Cohen, S., D. R. Sherrod, and M. S. Clark. 1986. "Social Skills and the Stress-Protective Role of Social Support." *Journal of Personality and Social Psychology* 50: 963-973.

Cook, J. 1983. "Citizen Response in a Neighborhood Under Threat." *American Journal of Community Psychology* 11: 459-471.

Coyne, J. C., C. Aldwin, and R. S. Lazarus. 1981. "Depression and Coping in Stressful Episodes." *Journal of Abnormal Psychology* 90: 439-447.

Davis-Brown, Karen and Sonya Salamon. 1988. "Farm Families in Crisis: An Application of Stress Theory to Farm Family Research." Pp. 47-55. in *Families in Rural America: Stress, Adaptation, and Revitalization*, edited by Ramona Marotz-Baden, Charles B. Hennon, and Timothy H. Brubaker. St. Paul, MN: National Council on Family Relations.

DeLongis, Anita, James C. Coyne, Gayle Dakof, Susan Folkman, and Richard S. Lazarus. 1982. "Relationship of Daily Hassles, Uplifts, and Major Life Events to Health Status." *Health Psychology* 1: 119-136.

Duncan, Stephen F., Robert J. Volk and Robert A. Lewis. 1988. "The Influence of Financial Stressors Upon Farm Husbands' and Wives' Well-Being and Family Life Satisfaction." Pp. 32-39 in *Families in Rural America: Stress,*

Adaptation, and Revitalization, edited by Ramona Marotz-Baden, Charles B. Hennon, and Timothy H. Brubaker. St. Paul, MN: National Council on Family Relations.

Eckenrode, J. 1983. "The Mobilization of Social Supports: Some Individual Constraints." *American Journal of Community Psychology* 11: 509-528.

Edelstein, Michael R. 1988. *Contaminated Communities: The Social and Psychological Impacts of Residential Toxic Exposure*. Boulder: Westview.

Edelstein, Michael R. 1989. "Psychosocial Impacts on Trial: The Case of Hazardous Waste Disposal." Pp. 153-177 in *Psychosocial Effects of Hazardous Toxic Waste Disposal on Communities*, edited by Dennis L. Peck. Springfield, IL: Thomas.

Edelstein, Michael R. and Abraham Wandersman. 1987. "Community Dynamics in Coping with Toxic Exposure." *Neighborhood and Community Environments*, edited by Irwin Altman and Abraham Wandersman. New York: Plenum.

Fiore, J., J. Becker and D. B. Coppel. 1983. "Social Network Interactions: A Buffer or a Stress?" *American Journal of Community Psychology* 11: 423-439.

Fischer, C. S. 1982. *To Dwell Among Friends: Personal Networks in Town and City*. Chicago: University of Chicago Press.

Folkman, S. and R. S. Lazarus. 1980. "An Analysis of Coping in a Middle-Aged Community Sample." *Journal of Health and Social Behavior* 21: 219-239.

Folkman, S., R. S. Lazarus, C. Dunkel-Schetter, A. DeLongis, and R. J. Gruen. 1986. "Dynamics of a Stressful Encounter: Cognitive Appraisal, Coping, and Encounter Outcomes." *Journal of Personality and Social Psychology* 50: 992-1003.

Freudenburg, William R. 1982. "The Impacts of Rapid Growth on the Social and Personal Well-Being of Local Community Residents." Pp. 137-170 in *Coping With Rapid Growth in Rural Communities*, edited by Bruce A. Weber and Robert E. Howell. Boulder: Westview.

Gold, Philip W., Frederick K. Goodwin, and George P. Chrousos. 1988. "Clinical and Biochemical Manifestations of Depression: Relation to the Neurobiology of Stress" (second of two parts). *New England Journal of Medicine* 319 (7, Aug 18): 413-420.

Hammen, C. L., S. E. Krantz and S. D. Cochran. 1981. "Relationship Between Depression and Causal Attributions About Stressful Life Events." *Cognitive Therapy and Research* 5: 351-358.

Haney, Wava and Jane B. Knowles, eds. 1988. *Women and Farming: Changing Roles, Changing Structures*. Boulder: Westview.

Hass, Gilda. 1985. *Plant-Closures: Myths, Realities and Responses*. Boston: South End Press.

Hass, Gilda. 1985. "Social Networks and the Coping Process: Creating Person Communities." *Social Networks and Social Support*, edited by B.H. Gottlieb. Beverly Hills, CA: Sage.

Jung, J. 1984. "Social Support and Its Relation to Health: A Critical Evaluation." *Basic and Applied Social Psychology* 5: 143-169.

Kasl, Stanislav V. and Sidney Cobb. 1979. "Some Mental Health Consequences of Plant Closings and Job Loss." Pp. 255-299 in *Mental Health and the Economy*, edited by Louis A. Ferman and Jeanne P. Gordus. Kalamazoo, MI: W. E. Upjohn Institute for Employment Research.

Keating, Norah C. 1987. "Reducing Stress of Farm Men and Women." *Family Relations* 36 (4): 358-363.

Keating, Norah C., Brenda Munro, and Maryanne Doherty. 1988. "Psychosomatic Stress Among Farm Men and Women." Pp. 64-73 in *Families in Rural America: Stress, Adaptation, and Revitalization*, edited by Ramona Marotz-Baden, Charles B. Hennon, and Timothy H. Brubaker. St. Paul, MN: National Council on Family Relations.

Kessler, Ronald C. and Jane D. McLeod. 1984. "Sex Differences in Vulnerability to Undesirable Life Events." *American Sociological Review* 49: 620-631.

Kessler, Ronald C. and Jane D. McLeod. 1985. "Social Support and Mental Health in Community Samples." Pp. 219-240 in *Social Support and Health*, edited by Sheldon Cohen and S. Leonard Syme. New York: Academic Press.

Kinichi, Angelo J. 1985. "Personal Consequences of Plant Closings: A Model and Preliminary Test." *Human Relations* 38: 197-212.

Kobasa, S. C. and M. C. Puccetti. 1983. "Personality and Social Resources in Stress Resistance." *Journal of Personality and Social Psychology* 45: 839-850.

Krause, N. and S. Stryker. 1984. "Stress and Well-Being: The Buffering Role of Locus of Control Beliefs." *Social Science and Medicine* 18: 783-790.

Lazarus, Richard S. 1981. "The Stress and Coping Paradigm." *Theoretical Basis of Psychopathology*, C. Eisedorfer, D. Cohen, A. Klienman, and P. Maxim, eds. New York: Spectrum.

Lazarus, Richard S. and Susan Folkman. 1984. *Stress, Appraisal, and Coping*. New York: Springer.

Lefcourt, H. M., R. A. Martin, and W. E. Saleh. 1984. "Locus of Control and Social Support: Interactive Moderators of Stress." *Journal of Personality and Social Psychology* 47: 378-389.

Lobao, Linda and Katherine Meyer. 1990. *Farm Family Adaptations to Severe Economic Distress: Ohio*. Ames, IA: North Central Regional Center for Rural Development.

Meyer, Katherine and Linda Lobao. 1991. "Consumption Patterns, Hardship and Stress Among Farm Households." *Research in Rural Sociology and Development V*, edited by Daniel C. Clay and Harry K. Schwarzweller. Greenwich, CT: JAI Press.

Mick, Stephen S. 1975. "Social and Personal Costs of Plant Shutdowns." *Industrial Relations* 14: 203-208.

Miller, S. M. and M. L. Green. 1985. "Coping with Stress and Frustration: Origins, Nature, and Development." *Socialization Of Emotions*, edited by M. Lewis and C. Saarni. New York: Plenum.

Mirowsky, John and Catherine Ross. 1986. "Social Patterns of Distress." Pp. 23-45 in *Annual Review of Sociology*, edited by Ralph H. Turner and James F. Short. Palo Alto, CA: Annual Reviews.

Nelson, D. W. and L. H. Cohen. 1983. "Locus of Control and Control Perceptions and the Relationship Between Life Stress and Psychological Disorder." *American Journal of Community Psychology* 11: 705-722.

Nelson, D. W. and L. H. Cohen. 1980. "The Life Cycle and Life Strains." *Sociological Theory and Research: A Critical Approach*, edited by H. M. Blalock, Jr. New York: Free Press.

Nelson, D. W. and L. H. Cohen. 1982. "The Social Contexts of Stress." *Handbook of Stress: Theoretical and Clinical Aspects*, edited by L. Goldberger and S. Breznitz. New York: Free Press.

Pearlin, L. I. and C. S. Aneshensel. 1986. "Coping and Social Supports: Their Functions and Applications." *Applications of Social Science to Clinical Medicine and Health Policy*, edited by L. Aiken and D. Mechanic. New Brunswick, NJ: Rutgers University Press.

Pearlin, L. I. and M. A. Lieberman. 1979. "Social Sources of Emotional Distress." *Research in Community and Mental Health*, Vol. 1, edited by R. G. Simmons. Greenwich, CT: JAI Press.

Perrucci, Carolyn C. and Dena B. Targ. 1988. "Effects of a Plant Closing on Marriage and Family Life." Pp. 55-72 in *Redundancy, Layoffs and Plant Closures: The Social Impact*, edited by R. M. Lee. Beckenham, Kent, U.K.: Croom Helm Ltd.

Quarantelli, E. L. and Russell Dynes. 1977. "Response to Social Crisis and Disaster." *Annual Review of Sociology* 3: 23-49.

Ritter, Christian, D. E. Benson, and Clint Synder. 1990. "Belief in a Just World and Depression." *Sociological Perspectives* 33: 235-252.

Rosenblat, Paul C. and Linda Olson Keller. 1983. "Economic Vulnerability and Economic Stress in Farm Couples." *Family Relations* 32 (October): 567-573.

Ross, Catherine E. and John Mirowsky. 1989. "Explaining the Social Patterns of Depression: Control and Problem-Solving or Support and Talking." *Journal of Health and Social Behavior* 30: 206-220.

Sandler, I. N. and B. Lakey. 1982. "Locus of Control as a Stress Moderator: The Role of Control Perceptions and Social Support." *American Journal of Community Psychology* 10: 65-80.

Snyder, Kay A. and Thomas C. Nowak. 1984. "Job Loss and Demoralization: Do Women Fare Better Than Men?" *International Journal of Mental Health* 13: 92-106.

Staten, Jay. 1987. *The Embattled Farmer*. Golden, CO: Fulcrum.

Stone, A. A. and J. M. Neale. 1984. "New Measure of Daily Coping: Development and Preliminary Results." *Journal of Personality and Social Psychology* 46: 892-906.

Thoits, P. A. 1981. "Undesirable Life Events and Psychophysiological Distress: A Problem of Operational Confounding." *American Sociological Review* 46: 96-109.

Thoits, P. A. 1982. "Conceptual, Methodological, and Theoretical Problems in Studying Social Support as a Buffer Against Life Stress." *Journal of Health of Health and Social Behavior* 23: 145-159.

Thoits, P. A. 1985. "Gender and Health: An Update on Hypotheses and Evidence." *Journal of Health and Social Behavior* 26: 156-182.

Tigges, Leann M. and Rachel A. Rosenfeld. 1987. "Independent Farming: Correlates and Consequences for Women and Men." *Rural Sociology* 52 (3): 345-364.

Vyner, Henry. 1984. "The Psychological Effects of Invisible Environmental Contaminants." *Social Impact Assessment* (July-September): 93-95.

Walker, Lilly and James L. Schubert. 1988. "Stressors and Symptoms Predictive of Distress in Farmers." Pp. 56-63 in *Families in Rural America: Stress, Adaptation, and Revitalization*, edited by Ramona Marotz-Baden, Charles B. Hennon, and Timothy H. Brubaker. St. Paul, MN: National Council on Family Relations.

Warr, P. and G. Parry. 1982. "Paid Employment and Women's Psychological Well-Being." *Psychological Bulletin* 91: 498-516.

Warr, P. and G. Parry. 1985. "Gender and Depression." *Trends in Neurosciences* 9: 416-420.

Wethington, E. and R. C. Kessler. 1986. "Perceived Support, Received Support, and Adjustment to Stressful Life Events." *Journal of Health and Social Behavior* 27: 78-89.

Whatmore, Sarah. 1991. *Farming Women: Gender, Work and Family Enterprise*. London: Macmillan.

Wheaton, B. 1982. "A Comparison of the Moderating Effects of Personal Coping Resources in the Impact of Exposure to Stress in Two Groups." *Journal of Community Psychology* 10: 293-311.

Wheaton, B. 1985. "Models for the Stress-Buffering Functions of Coping Resources." *Journal of Health and Social Behavior* 26: 252-264.

Williams, D. G. 1985. "Gender Differences in Interpersonal Relationships and Well-Being." *Research in Sociology of Education and Socialization*, edited by A. Kerckhoff. Greenwich, CT: JAI Press.

PART THREE

Community and Social Interaction

Introduction
Paul Lasley

Section three extends previous analyses of the farm crisis to the impacts that lie beyond the farm gate. As the farm crisis increasingly spread across the landscape, engulfing rural communities and altering social relationships, it became viewed as a rural crisis. The focus of this section is to explicate some of the tertiary impacts of the crisis on communities and how it has altered the actions of farm people within their communities. Others have termed the tertiary impacts, as the ripple effect, that wave action produced when a stone is thrown into a calm pool of water. While previous chapters have focused on farm families' perceptions and responses to economic hardship, chapters in this section explore how farm families view their communities and what new patterns of interaction have emerged. One of the neglected areas of inquiry has been to link economic hardship and emotional distress with community attachment and interaction.

While much secondary data exists on the widening gaps in social and economic conditions between metropolitan and nonmetropolitan counties, much less is known about how farm families perceive the impacts of economic decline within their own communities. Biere, in Chapter 9, explores farmers' assessment of how their communities had changed as a result of prolonged financial stress. As farm families have tightened their belts and hunkered down to weather the financial storm, rural business and social institutions were also forced to adapt. The stagnating rural conditions of diminished retail sales, shrinking tax base, outmigration, and contraction of the local economic activity gave rise to restructuring in many communities. Some of the common responses within communities have been to diversify to other industries, while others have been forced to consolidate social institutions, such as schools, and to reduce funding for public services. While many of these adaptations were fiscally prudent and necessary, in some cases they have reduced the overall satisfaction of community services and facilities, and in the most extreme cases represent a threat to the future viability of the community. In some cases, it appears that communities have actually improved the level of services to local residents and have made adjustments that have actually strengthened the commu-nity. The diversity of the local economy, county population, and proximity to a metropolitan area is shown to exert an important influ-ence on the viability of the rural communities and how respondents evaluated their community.

In Chapter 10, Lobao extends the discussion of farm crisis impacts by considering the political and organization involvement of farm families that were caught in the most pervasive rural restructuring decade since the 1930s. There are two schools of thought concerning farm residents' participation during times of economic decline. One school of thought is that farm families were expected to withdraw from social and political activities and the opposing view is that participation rates would increase. Lobao finds that sociopolitical involvement is positively related to one's socioeconomic status, suggesting that financially secure, affluent farmers were more likely to be involved. Contrary to the public image of widespread farm protests occurring during the farm crisis years, the data suggest that participation rates did not greatly increase. Indeed, the widening gaps between the fortunes of farm families is reflected in participation rates. Financially pressed farmers, those who likely needed assistance the most, were more likely to have withdrawn from sociopolitical participation, while at the same time, affluent farmers increasingly turned to collective actions to make their voices heard. This suggests the formation of an underclass within agriculture, comprised of downward mobile farm families, facing economic hardship, personal and familial stress, that drops out of community life, and withdraws from interaction in the community.

In Chapter 11, the editors of this volume attempt to put the farm crisis and rural restructuring into a policy perspective. While only 2 percent of America's population lives on farms, this tends to underestimate their economic contribution to the viability of many small agricultural communities. This study describes that the farm crisis is part of a much larger set of forces that are reshaping the face of rural America, and in many ways the 1980s only accelerated long term historical trends. The failure to recognize the events of the 1980s as ushering in a broad scale restructuring of rural America suggests the need for more interaction between social scientists and policy makers. The vastly different life chances among farmers, the disparity of opportunities across rural communities, and the availability of public and private resources has worked together to produce greater inequalities among rural people and communities. This chapter suggests the legacy of the farm crisis will persist for many years, unless more attention from both social scientists and policy makers is given to intervention strategies and comprehensive rural development efforts.

As we pause to marvel at the great strides of our highly efficient and productive agriculture, this chapter calls attention to the unintended consequences of inconsistent agricultural and rural policies and needed directions to rebuild the North Central Region.

9

Community Change

Arlo Biere

The restructuring of farming during the 1980s had major effects on rural communities. Not all communities were affected equally, however. One part of the North Central Regional Farm Survey was designed to learn more about how rural communities were affected.

The survey, completed in February 1989, also asked operators to evaluate changes in their community's conditions over the previous five years, from 1984 through 1988. Three sets of survey questions dealt with the community: operators' evaluations of changes in community services, facilities, and economic conditions; how much regret operators and spouses expected to feel if they were forced to leave their community; and operators' and spouses' observations of changes in community solidarity.

Farm operators were asked to evaluate changes in the quality of schools, job opportunities, health care services, child care facilities, shopping facilities, police and fire protection, adult education opportunities, banking services, and entertainment and recreation opportunities as well as changes in farmers' condition, area agribusinesses, area lenders, and their own farms. For each question, a respondent was given the choice of the following responses: improved, remained the same, grew worse, uncertain, and not available. Because very few "uncertain" or "not available" responses were given, I dropped them from the analysis, and compared the percentage improved with the percentage worsened. Although others responded that conditions had remained the same, the difference between the "improved" and the "worsened" categories indicates the direction of change as seen by the respondents as a group. That is an equivalent to using a trinomial variable with values of -1, 0 and +1 for responses of worsened, remained the same and improved, respectively. Thus, the mean of the such trinomial variable will equal the difference in the frequencies of those who responded improved and those who responded worsened. Using the trinomial variable specification allows one to test for differences in location among distributions. The simple test of difference of two means can be used to test for statistical difference of location for two distributions. A multiple comparisons test allows for comparisons among more than two populations and is used here.

In addition, operators and spouses each were asked how sorry they would be to leave their community if they had to move for some reason. The possible responses were very sorry, somewhat sorry, a little sorry, and not at all sorry. I combined the first three response categories to find the percentage who would be at least a little sorry. Finally, operators and spouses were asked about the changes in the past five years in neighboring, in neighbors helping each other, and in things they have in common with others in the community. The possible choices were much better, somewhat better, remained the same, somewhat worse and much worse. I combined the first two and the last two choices; the resulting response categories were better, remained the same, and worse. I tabulated the responses for the whole region, using the weights for each state to represent the population properly as published by Lasley and Fellows (1990).

The results were summarized first by state. The states in which respondents thought most strongly that farming conditions had improved were not the same as the states in which respondents thought most strongly that community conditions had improved. I hypothesized that factors such as industry composition and remoteness of the community affected the change in community conditions. For example, rural communities that were less dependent on farming would be affected less, as would rural communities that were closest to metropolitan areas or had a larger population.

Although all 12 states of the North Central region are major producers of farm commodities, the relative importance of farming to each state's economy varies. For example, a smaller share of state income comes from farming in the more populous industrial states of the corn belt and the lakes regions, which have large non-farm sectors, particularly in manufacturing and services. Changes in rural community conditions during the five-year period appear to be related to the industrial composition of the community and to its proximity to concentrations of population.

State and Regional Comparisons

In Nebraska, South Dakota, Iowa, and Minnesota, more farm operators said their farm's financial condition had improved during the five-year period than said it had worsened. In Kansas, equal percentages of respondents reported improved and worsened conditions. In the other North Central states, larger proportions said they had worsened. In Michigan, North Dakota, and Wisconsin, I found the greatest differences between those who said conditions had worsened and those who said they had improved: 23, 24, and 16 percentage points, respectively.

Operators' evaluations of their own farms' financial conditions were compared by state, with the trends in farm income in each state (Table 9.1). In the four states in which more operators said their farm's condition had improved than said it had worsened, farm income was higher in 1988 than in 1984. In Kansas, where equal percentages of respondents replied "improved" and "worsened," farm income increased 60 percent from 1984 through 1988, although it declined slightly between 1987 and 1988. Illinois, Indiana, Michigan, Ohio, and North Dakota experienced declines in farm income; in those states, more operators responded that conditions had worsened. That was not the case in South Dakota and Missouri, where the response margin favored "worsened" although farm income increased over the five-year period. The same was true for Wisconsin, but farm income had fallen there in 1988.

TABLE 9.1 Farm Income by State, in Millions of Dollars

	1984	1985	1986	1987	1988
Illinois	1,433	1,740	1,706	1,541	1,197
Indiana	916	810	938	936	693
Iowa	1,587	2,064	2,734	2,661	2,039
Kansas	917	1,294	1,598	1,558	1,463
Michigan	649	673	685	745	527
Minnesota	1,397	1,505	2,033	2,052	1,598
Missouri	556	795	792	880	886
Nebraska	1,318	1,702	1,939	1,896	1,890
North Dakota	637	689	783	744	370
Ohio	1,002	979	877	891	753
South Dakota	603	666	858	977	777
Wisconsin	1,336	1,293	1,658	1,699	1,436

Source: U.S. Department of Commerce, Bureau of Economic Analysis, "Survey of Current Business", Vol. 67:8, pp. 50-52; Vol. 68:8, pp. 34-36; Vol. 70:8, pp. 32-34.

Operators also were asked to evaluate the general change in the financial condition of all farmers during those five years. In random sampling, the respondents should represent all operators. If that is true and if operators' perceptions are accurate, the responses concerning all farms should be similar to those about the operator's own farm. Operators' assessments of their own farms' financial conditions, however, were much more positive than their assessments of all farms. This findings may reflect the fact that only the operators of surviving farms were surveyed; the sample did not include farmers whose operations had failed during the five-year period. It also may indicate that the improvement in the conditions of surviving farms is the result of operators' adjustments to the worsening financial situation in farming or that operators are overestimating other farmers' financial difficulties.

In most states, a large percentage of farm operators said that the financial conditions of agribusinesses in their communities had worsened (for example, 70 percent in North Dakota). State-to-state variations in responses concerning agribusinesses were similar to those in responses about farmers' financial conditions. Nebraska, the state with the most positive response regarding the financial condition of operators' own farms, also had the most positive responses from operators concerning the financial conditions of agribusinesses in their communities. This result is expected because agribusinesses depend heavily on farming either as the market for their products or as the supplier of commodities to be marketed and processed by them.

Operators' views of the lenders' current financial condition were more positive than views of the conditions of farming or agribusinesses. Although agricultural credit may account for much of a lender's business in many rural communities, that business also may include commercial credit, personal credit, and other financial services; thus the lender is less dependent on the farm economy than is agribusiness in general. Also, financial regulations in force during the period may have protected lenders' financial condition or delayed the exposure of the weakness of those lenders.

Job opportunities are major determinants of the future of a community. Farm operators were asked to evaluate the changes in job opportunities in their communities over the five-year period. In all states except Indiana, more respondents said that opportunities had worsened than said they had improved. In the industrial states of Ohio, Michigan, and Wisconsin, the responses were nearly even, or slightly more respondents said they had worsened. In the other states, especially in the plains states, many more respondents said job opportunities had worsened.

In all states except Iowa and the plains states (North Dakota, South Dakota, Nebraska, and Kansas), more operators said shopping facilities and recreational and entertainment opportunities had improved than said they had worsened. Regarding lenders' financial condition, more operators replied that they had worsened only in North Dakota and Kansas.

Public services, such as police and fire protection and adult education, as compared to private businesses are less affected by external economic factors because they are supported mostly through locally generated tax revenues. In all states, more operators replied that those services had improved than said they had worsened.

In each state, the responses concerning health care services were divided about evenly between "improved" and "worsened."

Influence of Industrial Composition and Remoteness

The state-by-state differences in community conditions raised questions. In particular, community conditions such as job opportunities by state were not related to changes in farm financial conditions by state. To learn more from the survey, I made comparisons by the type of county in which the operator lived. To do so, I merged the survey data with other data on the counties of the North Central region—the rural-urban classification by Butler (1990) and Chanyalew's (1990) classification of county economic type.

Butler classified all counties into two major classes—metropolitan and rural—and all rural counties into six subclasses. The latter were based on the urban population in the county and on the county's proximity to a metropolitan county (adjacent, non-adjacent).

Chanyalew divided rural counties into six economic types: farming-dependent, manufacturing-dependent, trade-dependent, service industry-dependent, mining-dependent, and government-dependent. He modified the procedures of Hushak and Gyekye (1984) and of Ross and Green (1985), who used sectorial labor and proprietors' income generated as share of county income to classify counties. Chanyalew first classified each county according to the industrial composition of labor and proprietors' income: farm-dependent if more than 18 percent of that income came from farming, mining-dependent if more than 18 percent came from mining, manufacturing-dependent if more than 28 percent came from manufacturing, government-dependent if more than 23 percent came from government, trade-dependent if more than 23 percent came from trade, and service-dependent if more than 23 percent came from services. As a result of that classification, some counties belonged to more than one class and others were unclassified. He, then, placed counties appearing in more than one class in the sector contributing the highest share of income, and applied discriminant analysis to produce a classification that placed each non-metropolitan county into one of the six classes.

Using the merged data, I tabulated by county type the responses to each question concerning changes in community conditions and ran the Tukey-Kramer multiple comparisons tests for unequal cell numbers. The tabulated responses are shown on Table 9.2 and the results of the Tukey-Kramer tests over county economic types for the same questions are shown on Table 9.3. On Table 9.3 the numbers represent the difference between the percentage responding improved and the percentage responding worsened (trinomial mean expressed in percentage terms). The lines drawn below those numbers are used to indicate differences in means that are not statistically significant. Any two means not underscored by the same line are not significantly different (at the 5 percent level). Also shown are the calculated F statistics for no significance among means.

For all county types, including metropolitan counties, more operators replied that job opportunities had worsened than that they had improved (Table 9.2). Operators in mining-dependent and trade-dependent rural counties gave the most negative responses regarding five-year changes in local job opportunities, even more negative than for farming-dependent rural counties. The least negative responses regarding changes in job opportunity were made by operators living in manufacturing-dependent rural counties; these almost matched responses for metropolitan counties. Service-dependent counties were rated next best.

The statistical significance of the differences over county economic types of perceived changes in job opportunities was determined using the Tukey-Kramer multiple comparison test for unequal cell numbers. The mean response from operators in metropolitan counties is significantly different from the mean for any rural county type other than manufacturing-dependent and the mean for rural manufacturing-dependent counties is significantly different from all other rural county types except service-dependent. Service-dependent, farm-dependent, government-dependent and mining-dependent have means that are not statistically different from each other. However, mining-dependent is not significantly different from the others because there were so few responses from mining-dependent counties. The mean for farming-dependent counties is larger and significantly different from the mean for trade-dependent counties.

The implications of the perceived change in job opportunities

TABLE 9.2 Farm Operators' Responses to Changes in Community Conditions, 1984 through 1988, by County Economic Type

County Type	Jobs	Health Care	Child Care	Shopping	Police, Fire	Banks	Entertainment, Recreation
Metropolitan				percent of those responding			
Improved	29	26	22	61	28	33	28
Worse	31	15	6	12	4	13	10
I-W[1]	-2	11	16	49	24	20	18
Farming-dependent							
Improved	16	21	16	30	22	19	13
Worse	42	22	8	33	6	19	24
I-W	-26	-1	8	-3	16	0	-11
Mining-dependent							
Improved	13	20	12	34	18	27	10
Worse	65	25	8	29	12	17	20
I-W	-52	-5	4	5	6	10	-10
Trade-dependent							
Improved	11	20	17	28	21	20	13
Worse	54	23	13	39	5	17	31
I-W	-43	-3	4	-11	16	3	-18
Government-dependent							
Improved	13	21	13	39	25	21	17
Worse	43	26	10	22	6	17	23
I-W	-30	-5	3	17	19	4	-6
Service-dependent							
Improved	19	23	19	45	26	26	18
Worse	39	21	11	19	7	15	19
I-W	-20	2	8	26	19	11	-1
Manufacturing-dependent							
Improved	25	24	20	51	24	30	19
Worse	34	17	6	14	6	15	16
I-W	-9	7	14	37	18	15	3

[1] I-W is the difference between the percentage who responded "improved" and the percentage who responded "worsened."

TABLE 9.3 Multiple Comparisons Tests of Farm Operators' Responses to Community Changes by County Economic Type

Job Opportunities (F=20.91)						
Metro	Mfg	Serv	Farm	Govt	Trade	Mine
-2	-9	-20	-26	-30	-43	-52

Groupings: (Metro, Mfg); (Mfg, Serv); (Serv, Farm, Govt, Trade); (Mine)

Shopping Opportunities (F=52.67)						
Metro	Mfg	Serv	Govt	Mine	Farm	Trade
49	37	26	17	5	-3	-11

Groupings: (Metro); (Mfg, Serv); (Serv, Govt); (Mine, Farm, Trade)

Banking (F=8.74)						
Metro	Mfg	Serv	Mine	Govt	Trade	Farm
20	15	11	10	4	3	0

Groupings: (Metro, Mfg, Serv, Mine); (Serv, Mine, Govt, Trade, Farm)

Entertainment Opportunities (F=22.55)						
Metro	Mfg	Serv	Govt	Mine	Farm	Trade
18	3	-1	-6	-10	-11	-18

Groupings: (Metro); (Mfg, Serv, Govt, Mine, Farm); (Trade)

Health Care (F=4.67)						
Metro	Mfg	Serv	Farm	Trade	Mine	Govt
16	14	8	8	4	4	3

Groupings: (Metro, Mfg, Serv, Farm); (Serv, Farm, Trade, Mine, Govt)

Child Care (F=5.85)						
Metro	Mfg	Serv	Farm	Trade	Mine	Govt
16	14	8	8	4	4	3

Groupings: (Metro, Mfg, Serv, Farm); (Serv, Farm, Trade, Mine, Govt)

Police (F=2.84)						
Metro	Serv	Govt	Mfg	Farm	Trade	Mine
24	19	19	18	16	16*	6*

Groupings: (Metro, Serv, Govt, Mfg, Farm); with Trade and Mine included at that level (broken line indicated by *)

1. Numbers are sample means expressed in percentage terms.
2. A line is drawn to connect all means that are not significantly different.
3. * indicates that the mean should be included with other means tied together with a line at that level. The line is broken because of unequal cell numbers which affect the significance level of differences.

among county types are several. Changes in job opportunities in the manufacturing dependent counties were not significantly different from those in metropolitan counties. While rural manufacturing-dependent counties were hit by farm restructuring and manufacturing restructuring, it appears that those counties fared about as well as metropolitan counties. Trade-dependent counties and mining-dependent counties (when ignoring the problem of lack of observations for mining-dependent counties) experienced the greatest loss in job opportunities as perceived by the operators. Rural trade centers were big losers because of the farm restructuring and the restructuring of retail trade, e.g., Wal-Marts and changes in people's buying habits in the form of bypassing traditional trade centers for larger ones.

Shopping facilities appear to have improved in metropolitan counties, manufacturing-dependent counties, service-dependent counties, government-dependent counties, and mining-dependent counties. The mean for metropolitan counties was statistically different from that of all other counties. The mean for manufacturing-dependent counties differed from all county types but service-dependent while the means for service-dependent, government-dependent and mining-dependent counties were not significantly different from each other. Shopping opportunities on average worsened for trade-dependent and farming-dependent counties. The worsening of shopping facilities in trade-dependent counties may be the result of rural residents' increasingly bypassing traditional rural shopping areas for modern malls and discount centers in more populous communities. In farming-dependent counties, shopping may have worsened because out-migration and downsizing in agriculture had reduced local retail markets.

Entertainment and recreation opportunities were rated as improved in metropolitan and in manufacturing-dependent rural counties. For all other rural counties, the "worsened" responses outnumbered the "improved." The most negative evaluation was registered for the trade-dependent counties, in which 31 percent of respondents stated that such opportunities had worsened. The closing of movie theaters in small towns is an example of decline in entertainment opportunities. When looking at the statistical results, one sees metro counties differ from all others, and manufacturing-dependent counties differ from farm and trade dependent counties.

Banking services fared better than shopping facilities or entertainment and recreation opportunities. For all types of rural counties, the number responding that banking services had improved equaled or exceeded the number reporting that those services had worsened. As concerns banking, metropolitan counties which had the most positive response were significantly different from only trade-dependent and farming dependent counties. Banking regulation may have insulated those institutions from market forces that would have led to greater consolidation of rural banks. More recent deregulation of banking has already produced changes in rural banking not detected here.

According to the respondents, health care improved in the service-oriented and manufacturing-oriented rural counties and in metropolitan counties but worsened slightly in the farming-dependent, mining-dependent, trade-dependent, and government-dependent rural counties. However, the results among rural counties are not statistically significant. Child care and police and fire protection improved for all types of county, and the results showed very little statistical significance.

A number of conclusions can be drawn from the data. Job opportunities declined in all types of rural counties, but manufacturing-dependent counties were affected least. Mining-dependent and trade-dependent counties were hit hardest.

Public services, such as police and fire protection and adult education, generally improved. In addition to jobs, other market-oriented activities worsened in certain types of counties, for example, shopping facilities in farming-dependent and trade-dependent counties and entertainment and recreation facilities in all except manufacturing-dependent counties.

I hypothesized that community's population size is a determinant of changes in community conditions. Population base is reflected in the size of the community's labor supply, the size of retail markets made by consumers, and the level of general economic activity in the community. Those conditions also are affected by nearby populations and markets, especially metropolitan areas. County data prepared by Butler (1990) reflect those differences; she categorized counties as either metropolitan or rural and according to proximity to a metropolitan county (adjacent or not adjacent) and by population size.

The survey results for those county classifications are shown in Table 9.4. The responses from farm operators in metropolitan counties are shown first for comparison. Those are followed by the responses from counties adjacent to metropolitan counties and then by responses from non-adjacent counties. Table 9.5 show the results of the multiple comparison tests for the same data.

On the basis of the response rates, job opportunities have deteriorated more in rural counties than in metropolitan counties. Among counties adjacent to a metropolitan county, the job opportunities were worst for the counties with an urban population less than 2,500. For non-adjacent counties, the largest and the smallest rural counties

TABLE 9.4 Farm Operators' Responses to Changes in Community Conditions, 1984 through 1988, by Metropolitan Proximity and County Population

County Type	Jobs	Health Care	Child Care	Shopping	Police, Fire	Banks	Entertainment, Recreation
				percent of those responding			
Metropolitan							
Improved	29	26	22	61	28	33	28
Worse	31	15	6	12	4	13	10
I-W[1]	-2	11	16	49	24	20	18
Rural Counties Adjacent to Metropolitan Areas with Urban Population of:							
More than 20,000							
Improved	22	23	17	52	22	31	21
Worse	34	12	6	11	4	15	11
I-W	-12	11	11	41	18	16	10
2,500 to 20,000							
Improved	24	22	18	48	24	26	16
Worse	35	19	7	19	6	15	18
I-W	-11	3	11	29	18	11	-2
Less than 2,500							
Improved	15	22	19	37	26	29	18
Worse	35	23	5	30	5	13	22
I-W	-20	-1	14	7	21	16	-4
Rural Counties Not Adjacent to Metropolitan Areas with Urban Population of:							
More than 20,000							
Improved	12	27	24	51	28	23	17
Worse	53	19	5	14	7	21	19
I-W	-41	8	21	37	21	2	-2
2,500 to 20,000							
Improved	19	22	18	40	24	25	17
Worse	40	20	9	25	5	16	21
I-W	-21	2	9	15	19	9	-4
Less than 2,500							
Improved	11	21	14	20	18	15	10
Worse	52	26	11	40	8	21	32
I-W	-41	-5	3	-20	10	-6	-22

[1]I-W is the difference between the percentage who responded "improved" and the percentage who responded "worsened."

had the largest margins, indicating that job opportunities had worsened. For all three categories of population size, a larger margin responded that job opportunities had worsened in non-adjacent counties than in adjacent counties of the same size.

Shopping facilities improved in counties adjacent to metropolitan counties. The percentage saying that these facilities had improved was highest for the largest counties and declined with county size. For counties not adjacent to a metropolitan county, improvement in shopping facilities depended on the county's population. Counties with urban populations exceeding 20,000 had responses similar to those of counties equivalent in size but adjacent to a metropolitan county. In the smaller, non-adjacent counties, shopping facilities did not fare so well; in non-adjacent counties with urban populations of less than 2,500 more responded that shopping had worsened than that it improved.

Entertainment and recreation opportunities showed a margin in favor of "improved" over the five years only for counties adjacent to metropolitan areas and with an urban population of more than 20,000. At the other extreme was the non-adjacent county with less than 2,500 urban population. For the other categories, the margin favored "worsened" only slightly. The indications are that entertainment and recreation opportunities improved most in metropolitan counties and deteriorated more in the more rural counties.

Health care has improved more in counties with the largest populations and in those adjacent to metropolitan counties. Improvement in child care shows little relationship to a county's proximity to a metropolitan county. Improvement appears to be related positively to population size for the non-adjacent counties: counties of more than 20,000 urban population had the largest margin of "improved" over "worsened" responses.

Operators' responses concerning police and fire protection showed no relationship to county population or to metropolitan adjacency, an indication that those services were affected little by the restructuring in farming.

Banking services in non-adjacent counties with less than 2,500 urban population apparently had deteriorated during the five-year period. The mean is statistically different from the same for all categories except non-adjacent counties with 2,500 to 20,000 population. In the other categories, the margin favored some improvement in banking services.

The data show that for the factors affected strongly by market forces—jobs, shopping facilities, and entertainment—rural areas are facing ever worsening conditions in relation to metropolitan and more highly populated areas. For police protection and fire protection, two examples of public services, I found no relationship between change in

TABLE 9.5 Multiple Comparisons Tests of Farm Operators' Responses to Community Change by County Size and Proximity

Job Opportunities (F=20.38)						
Metro	MA	LA	SA	MN	LN	SN
-2	-11	-12	-20	-21	-41	-41

(Lines connect: Metro–SA; MN–SN)

Shopping Opportunities (F=52.26)						
Metro	LA	LN	MA	MN	SA	SN
49	41	37	29	15	7	-20

(Lines connect: Metro–MA; MN–SA)

Entertainment and Recreation Opportunities (F=24.86)						
Metro	LA	MA	LN	SA	MN	SN
18	10	-2	-2	-4	-4	-22

(Lines connect: Metro–LA; MA–MN)

Banking Services (F=10.29)						
Metro	SA	LA	MA	MN	LN	SN
20	16	16	11	9	2	-6

(Lines connect: Metro–MA; MN–SN)

Health Care Service (F=4.57)						
Metro	LA	LN	MA	MN	SA	SN
11	11	8	3	2	-1	-5

(Lines connect: Metro–LN; MA–SN)

Child Care (F=5.81)						
MN	Metro	SA	LA	MA	MN	SN
21	16	14	11	11	9	3

(Lines connect: MN–MA; Metro–SN)

Police and Fire Services (F=5.12)						
Metro	SA	LN	MN	MA	LA	SN
24	21	21	19	18	18	10

(Lines connect: Metro–LA; SN separate)

1. Numbers are sample means expressed in percentage terms.
2. A line is drawn to connect all means that are not significantly different.

Metro = > 50,000 population.
LA = >20,000 and adjacent to metro county
MA = > 2,500 and <20,000 and adjacent to metro county
SA = ≤ 2,500 and adjacent to metro county
LN = ≥ 20,000 population, and not adjacent
MN = > 2,500 and < 20,000, and not adjacent
SN = ≤ 2,500 and not adjacent

quality and population size or proximity to a metropolitan county. Improvement in banking services was related slightly to population and proximity. Although banking is a private business, regulation probably has muted some of the market forces that otherwise might have resulted in more consolidation of banking and in the loss of services in some rural communities. In the 1980s, market forces played a strong role in shaping private industry as a whole, not only farming. When economies of scale are exploited, one can expect the net effect in many instances to be a worsening of conditions for small communities.

Community Solidarity

To learn about the changes in community solidarity, the survey asked operators and spouses to indicate changes in neighboring, in helping neighbors with work, and in commonality of values. The results varied little over states; they showed that in general, people were doing less neighboring, including helping each other less. On average, however, operators and spouses believed that what they had in common with the rest of the people in the community had changed little. Although people said they were more independent, they did not believe that shared community values had declined.

Finally, both operators and spouses were asked how much regret they would feel if forced to move from their community. Eighty-three percent of the operators and 80 percent of the spouses said they would be at least somewhat sorry to leave; that response indicates a strong community attachment. Leaving one's rural community is not an easy undertaking.

Summary and Conclusions

The 1980s was a period of major adjustment not only for farming but also for rural communities. In many such communities, adjustment took the form of community decline but was not necessarily correlated with financial changes in farming. For example, operators in Nebraska, South Dakota, Iowa, and Minnesota generally said the financial conditions of their farms had improved from 1984 to 1988. On the other hand, operators in those states replied more frequently that job opportunities, shopping, and other community conditions had worsened than did operators in other states such as those in the lake states and the eastern corn belt; in those areas, farming conditions were reported less frequently to have improved. Public services such as police and fire protection appear to have been maintained or improved regardless of the state.

The responses based on county economic type provided a clearer picture. Traditionally, rural counties depended on extractive industries of farming and mining. Generally, farming-dependent and mining-dependent counties are so classified not because of the strength of those industries but because of the lack of manufacturing, services, trade centers or government activities in the county. Farming and mining are not growth industries, and new technologies save labor and resources; therefore the labor freed by those developments has few promising alternatives for employment in such counties. As a result, the underemployed and the unemployed frequently leave to find work.

Trade-dependent rural counties are facing declines for many of the same reasons. New technology in wholesaling and retailing allows firms to substitute capital for labor and to enjoy the benefits of large-scale operations. The lower prices, greater variety, and other factors present in new discount centers and shopping malls are pulling retail trade from traditional trade-dependent counties. Furthermore, the population declines in the surrounding farming and mining counties that made up the trade area for the traditional trade-dependent counties have contributed to the decline in the economic base for those trade centers.

Manufacturing also faced restructuring in the 1980s, but the manufacturing-dependent and service-dependent counties appear to have fared best. Because those types of counties tend to be concentrated in the eastern corn belt and the lakes regions, rural counties are doing better in those areas than in the plains.

Population is another factor. Generally, the larger the population of the county and the closer the county is to a metropolitan area, the better the chance that the financial conditions of the rural community have improved. Population is important because it determines the sizes of the labor market and the local retail markets. In larger markets, it is easier to make a suitable match between the job and the job seeker. Also, a larger labor market is better able to endure shocks. Similarly, the larger the community retail markets, the more efficiently the demands for jobs and labor can be met.

Rural communities in America were created when the extractive industries were more important to the economy and when travel and communication limited the geographic area of markets. Those conditions have changed radically and continue to change in favor of more urban communities. The adjustments facing many rural communities are fundamental and life-threatening.

Restructuring first affected farming and manufacturing,

particularly the skilled and semiskilled jobs. More recently, restructuring has occurred in the white-collar occupations and somewhat in public services. It appears that no segment of the economy will be spared the challenges, opportunities, and agonies of restructuring caused by the information revolution, global markets, and other changes that affect all economies. Adapting to those changes remains a major challenge.

Finally, it is becoming recognized more clearly that adaptation involves many autonomous changes on the part of all entities in society. Those changes probably are more important than the planned adjustments found in government policies. Thus, it is a challenge to provide the education and information that allow individuals to adapt most successfully.

Notes

1. Contribution 93-230-B, Kansas Agricultural Experiment Station, Kansas State University, Manhattan, KS 66506.

References

Butler, Margaret A. 1990. *Rural-Urban Continuum Codes for Metro and Nonmetro Counties*. Washington, DC: U.S. Department of Agriculture, Economic Research Service.

Chanyalew, Demese. 1990. "Industry Structure in Rural America: The Effect of Change in Industry Structure on Job Loss in the Rural Counties of the North Central Region." Unpublished doctoral dissertation. Manhattan: Kansas State University.

Hushak, Leroy J. and Agyapong B. Gyekye. 1984. "The Relationship between Economic and Demographic Measures and Employment/Specialization." *North Central Journal of Agricultural Economics* 6: 48-58.

Kramer, C. V. 1956. "Extension of Multiple Range Tests to Group Means With Unequal Numbers of Replications." *Biometrics* 12: 307-310.

Lasley, Paul and Jacqueline Fellows. 1990. *Farm Family Adaptations to Severe Economic Distress: Regional Summary*. Ames, IA: North Central Regional Center for Rural Development.

Ross, Peggy J. and Bernal L. Green. 1985. *Procedures for Developing a Policy-Oriented Classification of Nonmetropolitan Counties*. Washington, DC: U.S. Department of Agriculture, Economic Research Service, Economic Development Division.

10

Organizational, Community, and Political Involvement as Responses to Rural Restructuring

Linda M. Lobao

The restructuring of the farm sector during the 1980s reverberated through rural society. Popular as well as academic accounts, including those in this volume, portray disruptions in three dimensions of social life: farm, household, and community. How did farmers respond organizationally and politically to these changing conditions? To what extent did they mobilize in defense of farm, family, and community? Two perspectives emerge from the chronicles of farmers' activism during the crisis.

First, farmers became more active in organizations and their activism took a variety of forms. Popular accounts in films and other media during this period highlighted dramatic events such as protests over farm foreclosures, standoffs between farmers and local law officials, and the growth of organizations on the fringes of the political right and left. Farm women's heightened activism was underscored in 1985 by the Congressional testimony of Hollywood farm wives—Jane Fonda, Sissy Spacek, and Jessica Lange. This view also suggests that substantial support from the non-farm community was mobilized through grass-roots coalitions of farmers, labor unions, churches, and community activists and through media events such as Farm Aid concerts. At the level of formal politics, farmers, through their many organizations, were said to have prompted federal and state governments to respond to the farm crisis. The result was an array of policies and programs to stabilize finances and help farmers continue farming.

A second, contrasting perspective stresses farmers' inaction due to factors such as their lack of financial resources and social support, increased work role obligations, and social psychological state. For example, anecdotal evidence points to times when farmers could not afford gas to drive into town to attend meetings or pay membership dues in farm organizations (Davidson 1990). Increased farm and off-farm work to make ends meet left little time for political activism. Studies documenting the social psychological effects of the farm crisis show that farmers' objective economic distress translated into negative perceptions about personal life satisfaction and to increased stress (Belyea and Lobao 1990; Heffernan and Heffernan 1986; Lobao and Meyer 1991; National Mental Health Association 1988). Such inner-directed responses tend to reduce political and organizational participation (Peterson 1990). In contrast to the bucolic ideal of the farm community as "rallying around its own," farmers experiencing financial distress often reported being shunned by friends and neighbors and being isolated from community life (Heffernan and Heffernan 1986; Rosenblatt 1990).

The purpose of this chapter is to assess these common portrayals of farmers' activism by examining patterns of voluntary organizational membership and political participation as well as their correlates. First, I describe farmers' experiences of crisis and change in farm, family, and community, and discuss how these experiences may be related to membership and participation. Then, I use data from the North Central Regional Farm Survey to address several issues. What level of women's and men's membership in farm and community organizations existed in 1988 and did this level change relative to the pre-crisis period? To what extent did spouses' and operators' experiences with regard to farm, family, and community change affect their organizational, community, and political participation? In the final section, I focus on the characteristics of two groups of farmers who became particularly visible during the 1980s: those who engaged in protest and farm political action, and those involved in farm women's organizations, reflecting the growing autonomy of that segment of the farm population.

Farm, Household, and Community Change: Objective Experiences and Subjective Appraisal

Changes in farming in the 1980s could be observed along two lines: as a consequence of long-term structural trends dating from the postwar period and as a shorter-term cyclical downturn or "farm crisis." As is well-known and documented, farms have declined in number and expanded in size. An increasingly dualistic system has emerged, characterized by relative changes such as the increase in the number of small part-time farms, a decline in simple commodity units or traditional family farming, and an increase in the market share of the largest farms (Lobao 1990). Farmers also experienced the worst financial stress since the Depression era: from 8 to 12 percent of those who were farming in 1980 went bankrupt, were foreclosed, or experienced financial restructuring during the decade (Stam et al. 1991: 2). Moderate to large-scale, commercially operated farms were said to suffer the brunt of the crisis.

Farm change obviously is intertwined closely with household well-being and adaptations, particularly insofar as a household depends on farming for its livelihood. Farm characteristics such as sales generated, labor and capital requirements, commodity specialization, and debt levels influence household adaptations such as the division of labor and off-farm employment decisions as well as family quality of life (Johnson, Lasley, and Kettner 1991). Historically, family farmers have coped with bouts of low prices and other financial strains by cutting back on consumption levels, or "self-exploiting" (Friedmann 1978). In another typical strategy, family members seek non-farm employment, thereby broadening their role responsibilities. Although these strategies have enabled family farmers to survive in an agricultural system that has become increasingly concentrated and centralized, survival has often come at the cost of poor quality of life and high social psychological stress. Studies on the farm crisis, including those in this volume, show that during this period, farm families also made adjustments in consumer expenditures such as postponing purchases; changing food, shopping, or eating patterns; reducing utility use; and seeking off-farm employment.

Farming affects also the well-being of the rural community. The survival of moderate-sized, family-operated farms has been linked to better community economic and social conditions over time (Lobao 1990). Studies focusing on the crisis period have further established farm-community links, particularly for areas dependent upon farming (Murdock et al. 1988). Downturns in farming tend to send secondary shocks to employment and wages in local suppliers. Tertiary effects may include a reduction of the community tax base and social services as well as unemployment in other industries and services. For rural areas during the 1980s, the farm crisis was compounded by downturns in other traditionally rural industries such as non-durable manufacturing, timber, gas, and mining.

Changes in farming and in broader rural society not only affect quantitative or objective aspects of well-being but also cause shifts in subjective perceptions. They call into question conventional ideologies in which the farm, the farm family, and the rural community are viewed as the preferred way of life. They upset traditional norms and beliefs. For example, previous shakeouts in farming had jeopardized older, less well-educated, and less entrepreneurial farmers, providing at least an understandable rationale for failure. The 1980s farm crisis, however, affected younger, better-educated, innovative farmers—those most successful in earlier decades. Hard work and education no longer provided insulation against financial distress. Farmers who tried to stay in farming by investing more money and labor in it, found themselves "in a fundamental identity crisis in view of future prospects" (Pongratz 1990: 12). The goal of preserving the farm and its associated lifestyle no longer made sense in the face of ever-increasing pressures. Some studies also documented changes in the norms of the rural community: a breakdown in social networks, neighboring, and community solidarity. Community support for those in financial distress appeared to weaken; in some farming communities, "collective depression" replaced the popular activism and commitment of earlier eras (Heffernan and Heffernan 1986; Sundet and Mermelstein 1988).

Experiences of Change and Organizational, Community, and Political Involvement

How might farmers' experiences with changes in the farm, household, and community affect their participation in voluntary organizations and politics? First, these changes may reduce the resources available to farmers for participation. A large literature exists on the relationship between economic and related resources and organizational/political participation. Resource mobilization perspectives, for example, observe that because those in poorer socioeconomic positions possess less money, must devote more free time to physical survival, or are more isolated from networks of information, the opportunity costs of participation tend to be higher for them.

In addition, real or perceived changes in community, farm, or household well-being may engender social psychological outcomes that affect participation. Feelings of failure, dissatisfaction with life, and fear of future socioeconomic conditions, tend to make individuals withdraw from interaction with others. Withdrawal from political or organizational participation is part of this more generalized response (Peterson 1990). According to Peterson, Americans appear to be especially prone to such behavior because the "individualistic ethos" that predominates in the United States fosters the view that life problems are "personal—and not due to large social and political institutions" (1990: 64). Thus in times of trouble, Americans tend to attribute responsibility to themselves. They employ individual coping strategies and turn away from group responses such as political or organizational activism. People who have less social contact with those around them are also isolated from information networks. They are less likely to be aware of local organizations and to seek membership (Unger and Wandersman 1985).

Perceptions of community also affect participation. A greater sense of community and attachment to place tends to promote participation in local organizations, which in turn strengthens community sentiments (Unger and Wandersman 1985). The perception that others will help in the face of community problems is linked to membership in local groups, although this perception does not apply to those who eventually become leaders (Oliver 1984).

Negative experiences and appraisals of change thus seem to depress social activism overall, but in some cases such discontent has been mobilized successfully. Studies of plant closings and environmental disasters, for example, point to increases in activism and in political consciousness as ways to deal with collective problems and to heal social psychological wounds (Edelstein 1988; Perrucci et al. 1988). These responses, however, tend to occur in localized settings. They also depend on the presence of mobilizing organizations, such as unions and citizens' councils, and on specific community leaders.

Real or perceived dissatisfaction may play a role for the few who choose less conventional actions and organizations. According to longstanding Marxist tenet, class location and the accompanying material conditions and socializing experiences should predispose those on the lower rungs of the stratification hierarchy to protest and other nonquiescent forms political participation. Relative deprivation theory holds that when the gap between economic expectations and real conditions is largest, the propensity for radical political behavior increases (Bergmann 1990).

In sum, the previous studies suggest that better objective conditions and positive perceptions are related to greater participation in conventional organizations and forms of activism. In view of the general American pattern of depressed activism in the face of poor conditions and negative appraisals, we should expect relatively few people to be involved in organizations and forms of activism that challenge the status quo. To extrapolate to our sample, farmers who experience the brunt of economic change will not respond generally with increased activism. Rather, farmers who are located higher in the farm stratification system and who have better farm and household resources should be more likely to participate in conventional organizations and political activities. Similarly, those who perceive fewer detrimental changes to their farms, households, and community should be more active. Few farmers are expected to chose more unconventional forms of activism; those that do are likely to have negative experiences and perceptions.

Analysis

I use data from the regional survey to address three issues: farm women's and men's patterns of organizational membership before and during the crisis period; the relationship between farmers' experiences and appraisals of change and their organizational, community, and political involvement; and the characteristics of those farmers who participated in protest and political action groups and in farm women's organizations. As noted in previous chapters, the North Central Regional Farm Survey focuses on spouses and operators who survived the downturns of the 1980s.

Membership in Community and Farm Organizations

In the survey, spouses were requested to delineate their own and operators' membership status in a number of farm organizations and in local political bodies such as school boards or town councils. Spouses were asked whether they were members, former members, or never members of these groups. In order to compare current patterns of organizational membership with those of the pre-crisis period, I adjusted the sample slightly to be comparable with Rosenfeld's (1985). Rosenfeld's data are based on a national random sample of farm women and men, conducted in 1980. Because our data are based on a sample of spouses and operators (about 98 percent and 97 percent of whom are, respectively, women and men), I delete the remaining 2 to 3 percent of our sample to reflect farm women (spouses) and their husbands and then weight the sample by spouses' response.

In 1989, the most often reported memberships for both farm men and women were in traditional farm organizations, such as the Farm Bureau, the National Farmers Organization, and the National Farmers Union (Table 10.1). Nearly 40 percent of the farm women and half of the farm men were members of these groups. Commodity producers' associations and supply and marketing cooperatives were the next most

TABLE 10.1 Membership in Community and Farm Organizations

	Women			Men		
	NC Region		National*	NC Region		National*
	Member	Former Member	Member	Member	Former Member	Member
	---percent---			---percent---		
Any farm organization such as Farm Bureau, NFO, National Farmers Union	39	11	33	49	15	50
Any women's branches of general farm organization, such as Farm Bureau Women	7	5	7	3	3	Not Asked
Any commodity producers' associations such as the American Dairy Association or National Wheat Producers Association	11	4	8	22	8	17
Any women's branches such as the CattleWomen or the Wheathearts	3	2	6	2	1	Not Asked
Women's farm organizations such as Women for Agriculture, American Agri-Women, or Women Involved in Farm Economics	2	2	2	0	1	Not Asked
Farm political action groups such as state Family Farm Movement or national Save the Family Farm Coalition	1	1	Not Asked	1	1	Not Asked
Local governing board such as school board or town council	6	6	Not Asked	14	15	Not Asked
Marketing cooperative	10	1	8	19	6	20
Farm supply cooperative	18	3	13	34	6	29
Any listed farm organization	47	—	45	60	—	64

Source: Based on Rosenfeld, 1985

* *National* Member

frequently mentioned. Membership in the latter organizations tends to be more regionally specific: for example, corn belt farmers are less likely to be members of cooperatives than are lakes and plains farmers, and farmers in the lakes states are most likely to be members of commodity associations (Lasley and Fellows 1990). Only about 1 percent of the farmers belonged to political action groups.

As has been noted in other studies, women are somewhat less likely than men to be members of farm organizations (Rosenfeld 1985) and far less likely to be formal leaders (Miller and Neth 1988). Although our data do not permit us to determine the reasons for this difference, Rosenfeld (1985) argues that contributing factors include work-role specialization, with men's work centered more on the farm; organizational environments less welcoming to women; and age and life cycle effects related to women's responsibility for household reproduction. The gap between male and female membership is relatively small for general farm organizations, but nearly doubles for commodity associations and in marketing and supply cooperatives. These differential patterns reflect frequently cited findings about decision-making patterns of farm men and women. Men tend to have greater control over farm management decisions such as what commodity to produce, where to purchase production supplies, and where to market goods. Hence they have greater actual contact with these external markets, organizations, and enterprises (Rosenfeld 1985). The gap between farm women and men is smaller in other decision-making arenas such as control over household resources, including farmland. Anecdotal accounts also suggest that farm input and marketing organizations and enterprises have been less welcoming to women than general farm organizations (Sachs 1983). Farm women's multiple work roles, which leave little free time, have been considered a further barrier to political participation (Miller and Neth 1988: 367; see also, Rosenfeld 1985). For example, a case study of women leaders in farm organizations found that many perceived a continued lack of farm men's involvement in housework and child care during the crisis, in spite of increases in women's off-farm work (Meyer and Lobao 1993).

The popular attention given to farm women's activism during the crisis suggested that their organizational membership might have increased over time. In fact, women's membership in general farm organizations was slightly larger than in Rosenfeld's national sample. Only about 2 percent of women, however, belonged to independent women's farm organizations such as American Agri-Women and Women Involved in Farm Economics, essentially the same proportion found in 1980.

A comparison of our general findings with Rosenfeld's (1985) indicates little change in farmers' organizational membership over the

1980s. The only sign that the farm crisis may have depressed membership is that 15 percent of the men and 11 percent of the women were no longer members of general farm organizations. Our data do not allow us to ascertain why or precisely when their membership may have ceased. Because our cross-sectional findings correspond to Rosenfeld's (1985), farmers probably discontinued membership due to factors not directly related to the crisis, such as life-cycle stage. Another possibility is that North Central farmers may have had higher membership rates and that these declined to about the national average in 1989. This does not seem to be the case, however. Although Rosenfeld's findings are not disaggregated by region, her regression analyses show that organizational participation rates of North Central farmers did not differ significantly from those of other regions in 1980.

In sum, for the survivors of the crisis, the changes of the 1980s apparently did not significantly alter previous patterns of organizational membership or motivate many persons to join political action groups. In our sample, however, farmers' lives were touched differentially by the restructuring process. The following analyses examine more closely how experiences and perceptions of restructuring are related to organizational and other public participation.

Organizational and Political Participation and Its Correlates

On the basis of the previous literature, farmers with better resources and positive experiences and perceptions of change should be more likely than others to belong to organizations and to participate more in community and political life. Because the variables needed to analyze the topic were available only if both the spouse and the operator answered, this analysis is limited to the 2,857 couples with matching spouse and operator questionnaires and with complete data on essential variables. Responses have been weighted by the operator data. Spouses and operators are analyzed separately. This controls for gender and allows for the examination of potentially distinctive patterns for spouses and operators.

Four measures of organizational and political involvement are examined. Organizational membership is measured by summing the total number of organizations (shown in Table 10.1) for which membership was reported. On average, both the spouses and the operators belonged to one organization each.

Community work is measured by how often the respondent works with other people in the community "to solve local problems."

Responses ranged from 1, "not at all" to 4, "once a week." Most operators and spouses reported they worked about one or two times a year on local problems (Table 10.2).

Political activism is measured by summing responses to the following conventional political activities in the past five years: attended public meetings, talked or wrote to government officials, signed a petition, and became more active in political groups. These data were reported by spouses for themselves and for the operators. Affirmative responses were coded 2, and negative responses were coded 1. High scores on the index thus indicate high political activism. More than half of both operators and spouses reported attending meetings; nearly half had signed a petition (Table 10.2). Fewer than one-fifth reported an increase in political activism over the farm crisis period. Spouses differed from operators mainly in that they were somewhat less likely to attend public meetings and to contact officials. Nearly all activism among farmers was of the previously named conventional types. Fewer than 2 percent of spouses and operators participated in a "protest over farm foreclosures."

A final indicator, derived from Oliver (1984), measures potential unconventional political involvement. Spouses only were asked, "If a farmer in your community has his/her farm foreclosed and some of the community residents thought it was unfair and organized a protest,

TABLE 10.2 Community Work, Political Activism, and Protest

	Spouses	*Operators*
	----------percent----------	
Community Work		
Once a week	6	8
Once a month	19	24
1-2 times a year	43	38
Not at all	33	30
Political Activism		
Attended public meetings	51	69
Talked/wrote to officials	31	40
Signed a petition	47	47
More active in political groups	15	16
Response to Farm Protest		
Would help	21	Not Asked
Probably help	42	
Unsure	17	
Probably not help	14	
Would not help	6	

would you help out?" Responses ranged from 1, "would not help" to 5, "would help". More than half of the spouses said they would or probably would help (Table 10.2).

The following sets of factors can be considered as contributing to involvement in conventional organizational and political activities. First, the results of long-term changes in farming are evident in the structure of the present farming system. Indicators of farm structure include gross farm sales coded into five categories ranging from 1, less than $10,000 to 5, $250,000 or more; whether the respondent has an off-farm job (affirmative responses coded 1, negative, 2); the percentage of farm labor provided by the family; and percentage of income from farming. Farmers higher in the farm stratification system or those with greater sales, more use of hired labor, greater proportion of their income from farming and less off-farm work, should have greater involvement.

In addition, the restructuring of farming affects objective household adjustments and resources, as measured by several variables. Total household debt-to-asset ratio is coded into four categories ranging from 1, less than 10 percent, to 4, more than 70 percent. Family income is a 10-category scale. Number of household adjustments made in the past five years is a scale summing yes (1) and no (0) responses to 12 common adjustments such as using savings to meet living expenses, selling possessions or cashing in insurance, purchasing more items on credit, changing food shopping or eating patterns to save money, and postponing medical or dental care to save money. Farmers with higher debt-to-asset ratios, lower incomes, and more adjustments should have lower organizational and political participation.

Objective, sociodemographic indicators not related directly to the farm crisis are employed because they are correlated typically with public participation. These indicators are age, years of education, and number of children under age 5. As noted in previous chapters, variables measuring farm structure, household adjustment and finances, and social demographic factors were from the operators' survey.

Perceptions of farm, family, and community change were collected separately from the operators and the spouses. Farmers were asked about the likelihood that they would farm for the next five years, their finances compared to those of other farmers, satisfaction with farming, family finances for the past five years, and family's quality of life. Responses to each of these variables were coded on a scale ranging from 1, "have become much worse," to 5, "have become much better." Indicators of community perception include neighboring, a two-item index evaluating the quality of neighboring and of neighbors helping one another in the past five years; and community attachment, a two-item index assessing the respondent's commonality with others in the community and how sorry the respondent would feel if she or he had to move away. High scores indicate greater community attachment. Spouses only were asked about sources of social support: groups, clubs, and organizations to which they belong; people that could help in times of trouble; close relatives; and close friends. The number of sources for each category ranging from 1, (none) to 5, (ten or more) were totaled. High scores on this index thus indicate high levels of social support.

Results of the relationships for operators and spouses are presented respectively in Tables 10.3 and 10.4. Zero-order correlations, indicating the strength of association between two variables, are used to provide a general overview of the findings. Multivariate analyses using ordinary least-squares regression were also performed using the previous farm structure, household finances/adjustments, social-demographic, and perception variables as independent predictors. Variables that remain significant in the regression equations are shown and the direction of their relationship with the dependent participation variables is highlighted. It should be noted that most relationships in the regression analyses followed the direction of the zero-order correlations.[1]

Organizational membership. Farm operators with higher gross sales, no off-farm work, and a larger percentage of income from farming belong to a greater number of organizations, as indicated by the zero-order correlations (Table 10.3). Because most of the organizations represented were conventional farm groups, these findings reflect what is generally known about their constituents: typically, they are larger, full-time, commercially oriented farmers. In contrast to expectations, farmers with higher debt-to-asset ratios belonged to slightly more organizations. However, as this relationship becomes non-significant in the multiple regression analyses, it may be likely due to other farm structural attributes. For example, larger farms tend to have more debt but are better able to support it than are smaller farms. More highly educated and younger farm operators also were involved more in membership groups. Among the perceptual indicators, farmers who had more confidence about their ability to farm for the next five years, who believed their finances were better than others', and who had higher community attachment were more likely to be members. The relationships are similar generally for farm spouses with the exception of age (Table 10.4). Higher perceived social support also is related to membership. Although the relationships cited above are statistically significant, their small size indicates a low degree of association between experiences and perceptions of change and membership.

Multiple regression analyses allow for the delineation of more salient relationships. These analyses also indicate that better

TABLE 10.3 Organizational and Political Participation and Its Correlates (Operators)[1]

	Organizational Membership	Community Work	Political Activism
Farm Structure			
Gross sales	.344***(+)	.177***(+)	.247***(+)
Off-farm work	.159***	-.019(+)	.080***
Family labor	-.011	-.064***	-.056**
Percent income from farming	.235***(+)	.064**	.139***
Household Finances/Adjustments			
Debt-to-asset ratio	.103***	.047*	.111***
Family income	.045*	.100***	.022
Number of adjustments	-.131	-.007	.050**(+)
Social-Demographic			
Age	-.074***	-.071***	-.042*(+)
Education	.133***(+)	.192***(+)	.144***(+)
Number of children	.018	.010	.003
Perceptions: Farm, Family, Community			
Will farm for next 5 years	.083***	.095***	.044*
Finances compared to others	.081***(+)	.113***(+)	.083***(+)
Satisfaction with farming	.032	.084***	-.002
Family finances	.012	.044*	-.018(-)
Family quality of life	.019	.083***	.020
Neighboring	.003	.070***	.021
Community attachment	.080***(+)	.151***(+)	.057**(+)

[1]Zero order correlation coefficients are shown. Variables that remain statistically significant in multiple regression analyses are noted in parentheses with their appropriate direction of relationship.

* Significant at ≤ .05
** Significant at ≤ .01
*** Significant at ≤ .001

TABLE 10.4 Organizational and Political Participation and Its Correlates (Spouses)[1]

	Organizational Membership	Community Work	Political Activism	Farm Protest
Farm Structure				
Gross sales	.231***(+)	.144***(+)	.165***(+)	-.014
Off-farm work	.056**	.003	.020	-.026
Family labor	-.002	-.034	-.035	.005
Percent income from farming	.168***(+)	.055**	.119***	-.035
Household Finances/Adjustments				
Debt-to-asset ratio	.055**	.016	.044*	.037
Family income	.025	.044*	.038*	-.065***
Number of adjustments	-.010(+)	-.019(+)	.025(+)	.103***(+)
Social-Demographic				
Age	.007(+)	.059**(+)	-.026(+)	.050**
Education	.092***(+)	.162***(+)	.182***(+)	-.041*
Number of children	-.026	-.061**	-.044*	-.067**(-)
Perceptions: Farm, Family, Community				
Will farm for next 5 years	.062**	.068***	.030	-.075***(-)
Finances compared to others	.109***(+)	.086***(+)	.082***(+)	-.066***
Satisfaction with farming	.044*	.076***	.036	-.044*
Family finances	.007	-.004(-)	.002	-.093***
Family quality of life	.003	.010	.003	-.075***
Neighboring	-.024(-)	.048**	-.033(-)	.093***(+)
Community attachment	.043*	.163***(+)	.063***(+)	.109***(+)
Social support	.110***(+)	.237***(+)	.132***(+)	.121***(+)

[1]Zero order correlation coefficients are shown. Variables that remain statistically significant in multiple regression analyses are noted in parentheses with their appropriate direction of relationship.

* Significant at ≤ .05
** Significant at ≤ .01
*** Significant at ≤ .001

financial conditions and resources are related to organizational membership. Farm operators and spouses with larger farms, greater income from farming, higher education, and stronger financial condition relative to others are more likely to belong to a greater number of organizations. There is also some evidence that the correlates of organizational involvement vary for spouses, with household adjustments and life-cycle variables more important. Spouses who are older and thus less constrained by household reproductive responsibilities are more likely to participate (the correlation between age and number of children under age 5 is r=.50). Some analysts have also argued that women's domestic responsibilities sensitize them more to financial adjustments, thereby providing a catalyst for political action (Meyer and Lobao 1993; Miller and Neth 1988).

Community work. The correlates of community work are similar in direction and magnitude. Larger farm size, greater income from farming, greater use of hired labor, higher family income, and higher educational level are related to a slightly greater frequency of community work. Positive subjective perceptions of operators and spouses also tend to be related to more frequent community work. Greater community attachment and social support (for spouses) are the most important of these correlates. There is also some evidence that the presence of young children discourages women's involvement, as indicated by the small negative correlation in Table 10.4.

The regression analyses highlight the previous patterns of greater activism among farm operators and spouses with larger farms, higher education, and stronger financial condition relative to others. For spouses, life-cycle and household factors affect community work. Older women, those whose households have made more financial adjustments, and those who perceive deterioration of family finances relative to the past are more likely to be involved in community work.

Political activism. Farm operators from larger farms, having a greater proportion of their income from farming, less off-farm work, and using more hired relative to family labor are more likely to have engaged in conventional forms of political activism. These relationships follow the same direction for spouses. Higher education, better finances as compared to others, and greater community attachment are also related to greater activism for both operator and spouse. The results from the regression analyses follow patterns noted in previous models, with the exception that age and number of household financial adjustments also positively influence operators activism as well as spouses.

Farm protest. Whether spouses would engage in protest over farm foreclosures depends little on the structural characteristics of their farms. Spouses whose households made more adjustments and had lower income were slightly more likely to endorse protest. The presence of young children was related to lower protest support. The relationship between perceptions and protest are small but follow a similar pattern. Negative perceptions of farm and family well-being are related to greater support for local protest initiatives, as are greater community attachment, more social support, and more favorable perceptions of neighboring. Many of these relationships remain significant in the regression analyses.

In sum, farmers' experiences and perceptions of change have only a small relationship to organizational membership, community work, and activism. Many of these relationships, however, tend to follow a consistent pattern. Farm operators and spouses who are on the higher rungs of the farm stratification system and who report better financial conditions relative to others are more likely to participate in farm and community organizations and conventional forms of activism. Perceived dissatisfaction with farm, family, and community tends to inhibit such involvement. These findings are similar to those of non-farm populations, in which participation is greater among the more affluent, more highly educated, and those more content with life opportunities (Peterson 1990). When one moves from the analysis of conventional political/organizational behavior to the potential for protest in the farm population, support for protest is increased by poor household economic well-being, including greater adjustments, but also by a positive feeling of community support. In general, the previous relationships held for both operators and spouses. However, spouse's participation was more closely related to household adjustments and life-cycle. In that a greater number of household financial adjustments were significantly related to spouses' organizational and political involvement, the classic contention that economic hardship may be mobilizing is supported but for women only. The relationships for age and number of children also provide some evidence that the household division of labor depresses women's activism.

Protest, Political Action, and Farm Women's Organizations

The farm crisis turned public attention especially to two groups of people: those whose dissatisfaction with farm change was expressed in protest and other collective action and farm women, who became more visible defenders of farm, family, and community life. Organizations appealing to the interests of these groups differed from conventional farm organizations, and especially from the Farm Bureau and commodity organizations, which typically dominated farm policy agendas. Farm political action groups such as the National Family Farm Coalition, Prairiefire Rural Action, the American Agriculture Movement, and the North American Farm Alliance advocate grassroots mobilization and alliances between farm and non-farm activists such as unions and church groups. Generally they take a "progressive" (liberal) stance. In contrast to the conventional farm and commodity groups, which organize clientele around existing farm support programs, these groups seek broader policy initiatives that range from farm credit and income issues to the environment and revitalization of rural areas. In their repertoire of political strategies, they tend to endorse protest when necessary. Farm women's organizations such as WIFE (Women Involved in Farm Economics), American Agri-Women, and Women for Agriculture provide a place for the independent organization of women. These organizations tend to focus on the dissemination of information to consumers and politicians, and they are generally non-partisan. Although only a small proportion of farmers in our study were involved in protest, political action, or women's organizations, some brief descriptions of their characteristics can be provided.

Of the 2,857 couples providing complete data, only 43 spouses and 55 operators participated in protest. Compared to the rest of the sample, these farm couples were significantly more likely to be members of farm and community organizations and to have engaged as well in a greater number of conventional political activities such as signing petitions and writing officials. These farmers operated more acres, had higher debts overall, and anticipated some problems in obtaining future loans. Their net family income was only slightly below that of the rest of the sample. Both operators and spouses had somewhat negative perceptions of farming; they reported lower overall satisfaction and greater likelihood of quitting in the next five years. In keeping with their behavior, spouses reported greater willingness to help out in cases of farm and community protest.

Although our study distinguishes farm political action groups from traditional farm organizations, we cannot delineate important variations among this latter group. The Farm Bureau, for example, tends to take a politically conservative, Republican stance (Browne and Cigler 1990). It is the largest general farm organization in terms of membership. Farmers often join the Farm Bureau because it provides insurance and other benefits rather than purely for its political perspective. The National Farmers Organization (NFO) began as a protest organization. The Farmers Union was established to support small family farmers and to oppose the extension of corporate farming. Both of these groups tend to support Democratic candidates and political initiatives. Within the NFO and the Farmers Union, local chapters may vary in goals and forms of action, such that they may be more similar to a farm political action group than to the Farm Bureau. Thus, although only about 1 percent of our sample belonged to farm political action groups, many more may have been members of chapters of traditional groups that supported similar goals. Thirty-six operators and 25 spouses, a total of 41 couples, belonged to farm political action organizations. Their characteristics were similar to those of persons who had participated in protest, except that they tended to be younger. They were more likely than the rest of the sample to be members of other farm organizations, to operate larger farms, and to receive less of their income from non-farm sources. They had higher overall debts, significantly lower family incomes, and expressed greater dissatisfaction with farming.

Farm women's organizations also show variations in political stances and strategies as well as differences within specific organizations by chapter. Overall, the goals and strategies of women's organizations are closer to those of conventional farm organizations, particularly the Farm Bureau and commodity groups. Forty-three spouses and two operators were members of farm women's organizations. Like the members of political action groups, these farmers belonged to a greater number of farm organizations; they also tended to operate farms with larger acreages and to depend more heavily on farm income. In contrast to members of political action groups, however, their objective well-being and their perceptions of farm change were positive. Members of farm women's organizations had more assets and fewer debts than the rest of the sample, reported better financial situations, and were more likely to recommend farming as a career. These women also were somewhat older on average and thus less likely to have young children.

In sum, farmers involved in protest, political action and farm women's groups are similar in that they tend to reflect profit-oriented family labor operations with households heavily dependent on farm income. Farmers involved in protest and political action, however, have experienced the downside of farm change whereas the positive experiences of those in farm women's organizations enable them to act as boosters for agriculture.

Summary and Conclusions

The purpose of this chapter was to examine farmers' organizational and political response to the restructuring of rural society in the late

1980s. In contrast to the belief that farmers as a group became more politically active, rates and patterns of activism showed little variation from those of pre-crisis years. There was some support for the view that farmers with positive experiences and perceptions of change were more likely to engage in conventional forms of activism. They were more likely to be members of farm and community groups, to have helped more with community problems, and to have participated in a greater number of political activities. Although these relationships were small, they tended to follow a pattern consistent with those of studies of non-farm populations: they emphasized greater sociopolitical involvement among individuals of higher socioeconomic status. Farm spouses and operators with larger farms, a higher proportion of their income from farming, higher education, better finances relative to others, as well as positive perceptions of community, tended to be more strongly involved. In the picture that emerges, entrepreneurial, full-time, more affluent farmers, integrated positively into rural life, were the major political and organizational participants during the farm crisis period.

In contrast to accounts of outbreaks of militancy among the farm population, fewer than 2 percent of the sample participated in protests; only about 1 percent belonged to farm political action groups. Although behavioral expressions of militancy were minuscule, a majority of farm spouses supported in principle the need for non-conventional collective behavior or protest; thus they seemed to recognize the seriousness of the crisis situation. Few farmers may have participated in actual protest because such events were localized and most farmers were not exposed to them. Farmers who supported protest both in principle and in behavior were more likely to have poorer household finances and more negative perceptions of farm and family well-being. Farm structural characteristics did not distinguish those who endorsed protest in principle from other farmers, but they did affect protest behavior. The few who engaged in protest were more dependent on farming for their livelihood, farmed more acres, and had more debt.

On balance, our findings suggest that the response among farmers with poorer financial conditions and negative perceptions of rural restructuring tended mainly to be withdrawal from organizational, community, and political activism. The few activists who responded with militancy resemble in some sense the farmer typified by the popular media of the period: the entrepreneurial, full-time family farmer beset by financial hardship and by the possible passing of a way of life. That economic hardship can be politically mobilizing, however, is given some support in the case of farm women's response to household adjustments.

These findings must be tempered by the limitations of the data. First, because the study is limited to survivors, it does not represent all of those who farmed during the crisis period. Thus it probably underestimates gross declines in membership in conventional farm organizations and former farmers' move toward protest and farm political action groups. In a related vein, the effects of financial hardship on community and political participation would tend to be muted because our sample reflects those whose finances permitted farm operation in 1989. Another issue is that not all questions were asked of both partners because of data collection procedures. For this reason, support for protest and the effects of social support could not be examined for farm operators. A final limitation is the lack of detailed information about types of farm organizations and farmers' specific activities in those organizations, such as frequency of attending meetings and leadership roles.

Why did we find little evidence of militancy and widespread mobilization among North Central farmers? First, as noted above, by focusing on survivors we tended to exclude the segment of the farm population most likely to be radicalized by the changes of the 1980s.

Second, farm households vary socioeconomically, and macroeconomic events such as the farm crisis do not affect them all in the same way. Farming has become increasingly differentiated internally over time so that most farmers can be considered as occupying "contradictory class" locations, neither fully proletarian nor fully capitalist. Farmers' political interests thus are heterogeneous (Mooney 1983). They are also shaped by the vagaries over time of an array of external forces, including fluctuations in prices for farm inputs and consumer goods, prices received for commodities, interest rates, land values, changes in international trade, and changes in local labor market opportunities. These forces in turn affect farm households differentially, depending on the household's class position, stage in family life cycle, non-farm labor market options, and so forth. Because economic downturns such as the farm crisis have highly uneven effects, farmers perceive their situation in different ways. Some view it as a "crisis," others as an opportunity; still others see little change. In the diversity of farmers' class locations, household conditions, and perceptions, the possibility of galvanizing issues becomes problematic.

Characteristics of the farm population also may have militated against wider collective response. Time, money, and social psychological resources would be least available for those in hardship. As other studies have noted, women would have faced special barriers due to work roles and responsibilities. Farm people's self-reliant, more independent life style may make them prone to self-blame when they are financially stressed (Heffernan and Heffernan 1986). The American tendency to blame the self rather than the system, and the consequent withdrawal from political action, may be greater in the farm population.

A final issue which cannot be explored in this study, is the role of farm organizations in mobilizing the farm population.[2] Our data show that larger, more prosperous, commercially oriented farmers were the major participants in organizations during the crisis period, as they have been throughout the postwar period. Possibly these individuals and their organizations did not define the situation as a fundamental crisis, one that could not be resolved through conventional political channels and by use of typical farm policies and programs. They may have acted to defuse rather than to mobilize popular discontent. In other words, the interests of the "affluent" farmer may have set the generalized tone of political response to the farm crisis.

Notes

1. Zero-order correlations among the farm structure, household finances/adjustments, social-demographic, and perceptions variables did not exceed moderate levels with the possible exception of some of the relationships among the perceptions variables. Even here, however, the highest correlation (between family finances and family quality of life) is only about .67 for both operators and spouses. Ordinary least-squares multiple regression analyses shows that about 10 percent of the variance in organizational membership, community work, and political activism, are explained by the correlates in Tables 10.3 and 10.4. This amount of variance is typical for studies on political behavior employing similar correlates (Peterson 1990).

2. Meyer and Lobao are studying the mobilizing strategies used by traditional farm organizations, farm political action groups, and women's farm organizations during the farm crisis period. Another view is that increased aid to the farm sector in the form of various state policies and programs diminished farmers' propensity to mobilize (Hsieh 1993).

References

Belyea, Michael J. and Linda M. Lobao. 1990. "The Psychosocial Consequences of Agricultural Transformation: The Farm Crisis and Depression." *Rural Sociology* 55 (Spring): 58-75.

Bergmann, Theodor. 1990. "Socioeconomic Situation and Perspectives of the Individual Peasant." *Sociologia Ruralis* 30 (1): 48-61.

Browne, William P. and Allan J. Cigler. 1990. *U.S. Agriculture Groups: Institutional Profiles*. New York: Greenwood.

Davidson, Osha Gray. 1990. *Broken Heartland: The Rise of America's Rural Ghetto*. New York: Anchor.

Edelstein, Michael R. 1988. *Contaminated Communities: The Social and Psychological Impacts of Residential Toxic Exposure*. Boulder: Westview.

Friedmann, Harriet. 1978. "World Market, State, and Family Farm: Social Bases of Household Production in an Era of Wage Labor." *Comparative Studies in Society and History* 20: 545-586.

Heffernan, William D. and Judith B. Heffernan. 1986. "The Farm Crisis and the Rural Community." Pp. 273-280 in *New Dimensions in Rural Policy: Building Upon Our Heritage*, edited by Dale Jahr, Jerry W. Johnson, and Ronald C. Wimberley. Washington, DC: U.S. Government Printing Office.

Hsieh, Horng-Chang. 1993. "Bonds That Tie and/or Divide: A Study on the Role of State Policy and Family Power Structure in Midwestern Farmers' Political Actions During the 1980s Farm Crisis." Ph.D. Dissertation. Columbus: Ohio State University, Department of Sociology.

Johnson, Gloria Jones, Paul Lasley, and Kevin A. Kettner. 1991. "Hardship and Adjustment among Farm Households in Iowa." Pp. 211-222 in *Research in Rural Sociology and Development: A Research Annual*, Vol. 5, edited by Harry K. Schwarzweller and Daniel Clay. Greenwich, CT: JAI. Press

Lasley, Paul and Jacqueline Fellows. 1990. *Farm Family Adaptations to Severe Economic Distress: Regional Summary*. Ames, IA: North Central Regional Center for Rural Development.

Lobao, Linda M. 1990. *Locality and Inequality: Farm and Industry Structure and Socioeconomic Conditions*. Albany: State University of New York.

Lobao, Linda M. and Katherine Meyer. 1991. "Farm Restructuring, Adaptations in Household Consumption, and Stress among Farm Men and Women." Pp. 191-209 in *Research in Rural Sociology and Development: A Research Annual*, Vol. 5, edited by Harry K. Schwarzweller and Daniel Clay, eds. Greenwich, CT: JAI Press.

Meyer, Katherine and Linda Lobao. 1993. "Engendering the Farm Crisis: Women's Political Response in the USA." Forthcoming in *Gender and Rurality*, edited by Sarah Whatmore. London: David Fulton.

Miller, Lorna Clancy and Mary Neth. 1988. "Farm Women in the Political Arena." Pp. 357-380 in *Women and Farming: Changing Roles, Changing Structures*, edited by Wava Haney and Jane B. Knowles. Boulder: Westview.

Mooney, Patrick. 1983. "Toward a Class Analysis of Midwestern Agriculture." *Rural Sociology* 48 (4): 563-584.

Murdock, Steve H., Lloyd B. Potter, Rita R. Hamm, Kenneth Backman, Don E. Albrecht, and F. Larry Leistritz. 1988. "The Implications of the Current Farm Crisis for Rural America." Pp. 141-168 in *The Farm Financial Crisis: Socioeconomic Dimensions and Implications for Producers and Rural Areas*, edited by Steve H. Murdock and F. Larry Leistritz. Boulder: Westview.

National Mental Health Association. 1988. *Report of the National Action Commission on the Mental Health of Rural Americans*. Alexandria, VA: National Mental Health Association.

Oliver, Pamela. 1984 "'If You Don't Do It, Nobody Else Will: Active and Token Contributors to Local Collective Action." *American Sociological Review* 49 (October): 601-610.

Perrucci, Carolyn C., Robert Perrucci, Dena B. Targ, and Harry R. Targ. 1988. *Plant Closings: International Context and Social Costs*. New York: Aldine.

Peterson, S. A. 1990. *Political Behavior: Patterns in Everyday Life*. Newbury Park, CA: Sage.

Pongratz, Hans. 1990. "Cultural Tradition and Social Change in Agriculture." *Sociologia Ruralis* 30 (1): 5-17.

Rosenblatt, Paul C. 1990. *Farming Is In Our Blood: Farm Families in Economic Crisis*. Ames: Iowa State University Press.

Rosenfeld, Rachel Ann. 1985. *Farm Women: Work, Farm, and Family in the United States*. Chapel Hill: University of North Carolina Press.

Sachs, Carolyn E. 1983. *The Invisible Farmers: Women in Agricultural Production*. Totowa, NJ: Rowman and Allenheld.

Stam, Jerome M., Steven R. Koenig, Susan Bentley, and H. Frederick Gale, Jr. 1991. *Farm Financial Stress, Farm Exits, and Public Sector Assistance to the Farm Sector in the 1980s*. Agricultural Economic Report Number 645. Washington, DC: U.S. Department of Agriculture, Economic Research Service.

Sundet, Paul and Joanne Mermelstein. 1988. "Community Development and the Rural Crisis: Problem-Strategy Fit." *Journal of the Community Development Society* 19 (2): 93-107.

Unger, Donald G. and Abraham Wandersman. 1985. "The Importance of Neighbors: The Social, Cognitive, and Affective Components of Neighboring." *American Journal of Community Psychology* 13 (2): 139-169.

11

Farm Crisis in the Midwest: Trends and Implications

F. Larry Leistritz and Katherine Meyer

In this volume, we and our collaborators have examined the effects of the restructuring of the farm sector in the Midwestern states, particularly the economic and social challenges that confronted farm operators and their families during the 1980s. We have focused on farming as a production system, on the well-being of farm households, and on the economy, services, and infrastructure of rural communities in the 12 states of the North Central region. This region is particularly significant nationally and internationally; it accounted for 41 percent of the nation's farms and 42 percent of total farm output in 1987. In addition, the region is dominated by family-scale farming operations (as opposed to large corporate farms), and almost two-thirds of its farmers report farming as their primary occupation. The North Central states also are reported to be among those hit hardest by the farm crisis of the 1980s (Leistritz and Murdock 1988), so the North Central Regional Farm Survey provides a unique opportunity to gain insights into the problems and changes that have become manifest and the ways in which farmers and communities have dealt with them.

The agricultural economists, sociologists, rural sociologists, and home economists who participated in this project have provided different, yet compatible, interpretations of data and unique, yet complementary, emphases in analysis. Agricultural economists focused on the financial characteristics of farm operations (Chapter 3). They assessed adjustments which operators and families made to reduce risks

(Chapter 4), and considered future strategies that farm operations and households will employ (Chapter 5). Economists detailed the farm management and spending patterns, the debt levels, the labor allocation, and the economic strategies and plans for the future found in Midwestern households as they dealt with the crisis. Sociologists examined gender changes in the division of labor and decision making within the household (Chapter 6). They researched community attachment, commitment to farming, and perceptions of the quality of rural life among farm operators and spouses (Chapter 7). In addition, sociologists examined mental health indicators of farm men and women (Chapter 8) and the political mobilization of farmers (Chapter 9). They also explored work and household changes, perceptions about life, and social psychological well-being.

From the joint labor of specialists in different disciplines, we can draw conclusions about farm operations, households, and communities. These conclusions center on the widening disparities within the rural sector, on common strategies and mechanisms for survival, and on economic and social profiles of farmers who prevailed during the 1980s.

Disparities Within the Rural Sector

One inescapable conclusion of our analysis is that the long-term trend toward fewer and larger farm production units continued unabated during the "farm crisis" of the mid-1980s. The survey results also point to a pattern of disparities among farms and farm households—disparities that many observers believe widened during the 1980s. These disparities were manifested in the survey data in a number of ways. For example, 42 percent of the households reported a total family income of less than $20,000, whereas almost 22 percent had total family incomes of $40,000 or more. Similarly, 12 percent of the families had debt-asset ratios exceeding 70 percent, whereas 30 percent had no debt.

The disparities indicated by the financial data were reflected as well in the respondents' perceptions. When asked how their family's quality of life had changed over the past five years, almost 40 percent said it had improved, but about one in six stated that it had become worse. In reporting how their satisfaction with farming had changed over the past five years, 24 percent of operators and 19 percent of spouses said it had improved, but 28 percent of operators and 30 percent of spouses said it had become worse.

We also observed differences in farmers' organizational and political responses to the events of the 1980s (Chapter 10). Contrary to the view that farmers as a whole became more politically active during that decade, the survey results show little variation in overall rates and patterns of activism, in comparison with the years before the crisis. The farmers who obtained a larger proportion of their income from farming, who operated larger farms, and who made greater use of hired labor tended to be more strongly involved. They were more likely than their counterparts with opposite characteristics to be members of farm and community groups, to have helped with community problems, and to have participated in a greater number of political activities. Higher family income and education and positive perceptions of farm, household, and community well-being also were related to greater involvement.

Disparities among counties adjacent to metro areas, manufacturing-dependent rural counties, and farm-dependent counties became exaggerated during the crisis period. When the net family income of farm households in metro counties was compared with that in non-metro counties, the metro county residents were represented more frequently in the higher income groups. About 26 percent of metro households reported a family income of $40,000 or more, compared with 21 percent in non-metro counties; 40 percent had incomes of $30,000 or more, compared with 35 percent in non-metro areas.

Farm households in metro counties (who made up about 22 percent of the sample) tended to be somewhat less dependent on the farm for their livelihood than were their counterparts in non-metro areas. The farm operation accounted for only 50 percent of total family income for the metro group, compared with 63 percent for non-metro residents. The remainder of the families' income was derived from off-farm employment and/or other non-farm sources (e.g., investment income). The greater reliance of metro residents on non-farm sources of income reflects both their greater propensity to work off the farm and the smaller size of typical farming operations in metro counties. Metro residents more often operated farms with gross sales of less than $40,000 (52 percent vs. 41 percent for non-metro respondents), and the farm operators reported more frequently that they worked off the farm (46 percent vs. 38 percent). We found little difference between metro and non-metro households, however, regarding the spouses' off-farm employment. About 50.4 percent of spouses in metro counties worked off the farm in 1988, compared with 49.3 percent in non-metro counties.

The survey respondents were queried about changes they had perceived in various aspects of their communities over the past five years. Job opportunities in the community were generally perceived as having grown worse. Regardless of the economic base of the county or its metro status, respondents reported that job opportunities had become poorer. On the other hand, respondents in each type of county believed that child care and police and fire protection had improved. Banking services also were believed to be better, except in farming-dependent counties (where equal percentages of respondents rated them better and worse) and in rural counties not adjacent to metro areas.

Other aspects of community services and conditions drew mixed responses. Health care was thought to be improved in metro counties as well as in rural counties dependent on manufacturing and services, but to be slightly worse in rural counties that depend primarily on farming, mining, trade, and government. Shopping was regarded as greatly improved by respondents who live in metro counties and in rural counties that depend on manufacturing, services, or government, and as slightly improved in mining-dependent counties. Shopping was viewed as worse in rural counties that are farm- or trade-dependent. Entertainment and recreation were seen as better in metro counties and in manufacturing-dependent rural counties, but as worse in all other types of rural counties.

Overall, rural communities in the North Central region have undergone a variety of changes, some caused by changes in the agricultural sector and some due to other causes. These changes in turn have affected the job opportunities and community services available to farm households and to other rural residents.

In sum, as economists and sociologists examined findings on farm operations, communities, and households in the 12 Midwestern states, it became obvious that the old disparities had widened and that new ones had emerged. Further, the effect of the economic conditions of the 1980s on farm households was evident in changes in the farm population. From 1980 to 1990 the farm population of the North Central region declined 34 percent. This decrease has implications for a variety of public services and private-sector activities in rural communities.

Strategies and Mechanisms for Surviving

Analyses of the Midwestern states confirmed and extended research from case studies and single states, which found Midwestern farmers changing, adjusting, adapting, strategizing, scrambling, and coping with adversity during the 1980s. Most of the region's farmers and ranchers made substantial adjustments in their farming operation in an effort to adapt to changing conditions. More than three-fourths of the farm and ranch operators in the survey reported giving more attention to marketing, and more than 70 percent reported that they had postponed a major farm purchase. More than 60 percent reported taking specific actions to reduce debt. These percentages were similar across subregions and across different levels of farm sales and net family income, an indication that most of the region's farm operators and their families felt the effects of the farm restructuring of the 1980s. The fact that many of these families have been able to continue in farming is a tribute to their resourcefulness and perseverance in the face of adversity.

The farm operators' responses to questions about changes they planned to make in the near future show that they have not forgotten the experiences of the 1980s. About two-thirds planned to pay closer attention to marketing, and more than half planned to keep more complete financial records and to further reduce debt. About one-fourth planned to diversify their farming operation by raising livestock; 15 percent wanted to add new crops. About 23 percent planned to seek off-farm employment, and about 22 percent planned to retire from farming or quit farming within the next five years. As producers attempt to implement these planned changes, the implications for both farm and non-farm sectors of rural communities may be substantial.

The adjustments in the farming operations of the North Central region during the mid-1980s were reflected in adjustments made by farm households. More than half of the households in the survey reported that they had postponed major purchases during the last five years; the proportion increased to 82 percent among the most highly leveraged farms. Almost half of the households reported that they had used savings to meet living expenses; this proportion increased to 71 percent in the most highly leveraged group. More than one-third of the operators and 38 percent of the spouses had taken off-farm employment; these proportions were 52 percent for operators and 62 percent for spouses in the most highly leveraged group.

Even after making the adjustments discussed above, many of the farm households in the region had rather limited financial resources. About 42 percent of the households reported a total family income of less than $20,000 (19 percent had less than $10,000), and about 12 percent had debts amounting to 70 percent or more of the total value of their assets. Many of these families seemed likely to face continuing difficulties in meeting household needs from their current income, and their leverage situation made them vulnerable to any unfavorable development (such as one poor crop year).

In 1987, two-thirds of the region's farms had gross sales of less than $50,000, a sales volume that is generally too small to provide an adequate level of family income. This fact indicates that most of the region's farm families either must generate much of their income from off-farm sources or must find a means to substantially enhance their farm income, either by expanding their production of current enterprises or perhaps through alternative enterprises. This situation emphasizes the interdependence of the farm and the non-farm sectors of the rural economy; many farm households depend heavily on earnings from non-farm employment. On average, in about 60 percent of the farm households represented in the survey, the operator, the spouse, or both were employed off the farm, and income from non-farm sources accounted for 40 percent of total family income.

Social and Economic Profile of Midwestern Farmers

Readers of this volume may have noted the emergence of profiles of farmers who survived in the 1980s in prosperity, farmers who barely managed, and farmers somewhere between those two extremes. Taking gross farm sales as an indicator of differentiation among production units, we examined those profiles empirically. The task was neither conceptually simple nor empirically perfect. For example, the debt-asset ratio, the amount of hired labor, the percentage of income from farming, and total family income would figure into an overall assessment of differentiation among production units. Yet, the creation of profiles using multiple indicators of prosperity would reduce the ability of other sets of variables to distinguish among categories of differentiation. Thus, we used gross farm sales as a proxy for general farm differentiation, although we were aware that our measure was not totally comprehensive.

Taking variables of interest in each chapter of this volume, we divided them into sets that could be distinguished conceptually. The farmers' demographic characteristics, the strategies they employed to survive, the context of community life and friendship networks, political and organizational activity, stress, and general satisfaction with farming were entered as separate sets into canonical discriminant analysis. Our task was to identify variables from each set which distinguished substantial farm sales ($500,000 or more) from those squeezing the lower margins ($40,000 or less). In fact, farmers between those two extremes formed a third group, larger than either of the other two. The largeness of that middle group somewhat violated the empirical requirement that the number of cases in each group be nearly equal. Yet, it was not sensible conceptually to further subdivide the middle group. For example, for what purpose would we distinguish those who were mostly but not clearly high volume operations from those who were mostly but not clearly in the part-time farm category? Thus, we maintained the large middle group and focused on the extremes in analysis to provide a concise profile.

According to the analysis, the farms with large gross sales fell into two categories based on the size of their farms. If the farm's acreage was small and if the male farmer was either over 65 or in his 20s, gross farm sales were low. (The spouse's age did not affect sales.) If the farm's agreage was large, all of the farmers' demographic characteristics (age, education, and number of children) were irrelevant; large farms had high sales volumes. The continued uneven development of the farm sector was underscored.

The demographic variables (age, education, number of young children) were the second most effective set of variables in discriminating among categories of gross farm sales. The linear combinations produced in this way—one emphasizing acreage and the other, the operators' age—had F coefficients of 89.61 and 4.65 respectively; the first was significant at the .0001 level and the second at the .01 level (see Table 11.1).

The organizational activity of farm operators (belonging to farm organizations and commodity producers' associations, and attending public meetings) was the set of variables that distinguished most accurately among categories of farm sales ($F=24.723$; $p=.0001$). In fact, in subsequent regression analysis, these variables indicating activism were among the strongest predictors of high gross farm sales. Membership in a marketing cooperative and in a commodity producers' association also differentiated farmers by gross farm sales ($F=4.96$; $p=.0002$).

Other sets of variables did not discriminate so clearly as did demographic variables and activism. Nonetheless, they advanced our task of presenting overall profiles of farms that prospered and farms that did not. In 42 percent of the cases, farms with high sales could be distinguished from the other two groups on the basis of farmers' plans to continue farming for the next five years and on whether they would recommend farming to others, including their children and relatives ($F=7.74$; $p=.0001$). Satisfaction with the family's quality of life and with farming also correctly distinguished different categories of gross farm sales ($F=5.11$; $p=.0001$). Because 58 percent of the cases were not distinguished by enthusiasm for farming, we cannot infer that most farms which were prospering were operated by enthusiastic farmers, while the opposite was true for those which were not. Rather, enthusiasm for farming corresponded to success in farming only for some respondents; for many others, it bore no relationship.

Plans for the next five years divided empirically into two categories: plans for marketing and economic practices. The intention to focus on marketing distinguished clearly the farms with high, medium, and low sales ($F=16.89$; $p=.0001$). So did intentions to reduce short-term

TABLE 11.1 Standardized Discriminant Coefficients of Selected Measures on Gross Farm Sales, by Set

Independent Variable	Function 1	Function 2
Demographic Variables Set		
Operator's Age	-.4534	1.6244
Spouse's Age	.1477	-.7191
Acreage	1.0359	.3427
F-Statistics	89.6143***	4.6504**
Organization Membership Set		
Operator in general farm organization	.4044	-.0427
Operator in commodity association	.4351	-.4365
Spouse in commodity association	-.0806	.3446
Spouse in women's commodity cooperative	.2422	-.6012
Spouse in marketing cooperatives	.1426	.6846
Operator participated in public meetings	.6113	.2064
F-Statistics	24.7237***	4.9611***
Future Strategies Set		
Pay closer attention to market	.9405	.2113
Reduce short-term debt	-.3134	.5601
Reduce expenditure in hiring extra labor	.0553	.8327
F-Statistics	16.8893***	4.6349**
Satisfaction Set		
Would recommend farming (operator)	.4394	.3308
Satisfied with quality of family life (operator)	-.0902	.5955
Satisfied with farming (operator)	-.1059	.4675
Would recommend farming (spouse)	.5502	.1649
Continue to farm in next 5 years (spouse)	.5944	-.5325
Financial situation better than others	-.0507	.3906
F-Statistics	7.7447***	5.1062***

continues

TABLE 11.1 (continued)

Independent Variable	Function 1	Function 2
Social Psychological Variable Set		
Personal level of stress (operator)	1.2298	-----[1]
Concern with stress (operator)	-.2705	-----[1]
Daily stress level (operator)	-.2150	-----[1]
Personal level of stress (spouse)	.3529	-----[1]
Daily stress level (spouse)	-.5010	-----[1]
F-Statistics	5.0422***	-----[1]
Community Connectedness		
Participated in solving community problem (operator)	-.4560	-----[1]
Neighboring over past 5 years (operator)	.1900	-----[1]
Number of community organizations in which member (spouse)	.8545	-----[1]
Number of social support	.4880	-----[1]
F-Statistics	10.3126***	-----[1]

[1]Not available because of insignificance
** $p < .05$
*** $p < .001$

debt and hired help as an expenditure (F = 4.63; p = .009). A variety of farmers, regardless of their gross farm sales, employed other plans such as diversifying, buying or renting more or fewer acres, retiring or leaving farming, retraining, making decisions to share, reduce, or postpone equipment costs, and implementing further accounting practices. Few of these plans distinguished farmers currently working farms with high, medium, or low sales. Farmers had different intentions for marketing, reducing short-term debt, and paying for hired help over the next five years, no matter where their farms stood economically at the time of the survey.

The amount of stress and of personal concern about stress distinguished farm men and women whose farms had different levels of gross sales (F=5.04; p=.0001). So did indicators of farmers' connectedness to their communities. Gross farm sales corresponded to respondents' working with others on local problems, experiencing a sense of neighboring, and belonging to a number of organizations (F=3.4; p=.01).

Canonical discriminant analysis, analysis of variance, and regression analysis all created a clear picture of farms with high gross sales. Such farms generally were large, and the couples living there belonged to farming organizations—both general farm organizations and commodity producers' associations. The farmers were publicly active in their communities. They attended public meetings, and many of them would recommend farming to their children or relatives. Many planned to reduce short-term debt and to pay close attention to marketing in the future.

Implications

The findings reviewed here have a number of implications in regard to changes in the structure of agriculture and the economic base of rural communities. In addition, the findings provide insights into needs for future research.

The findings of this study support those of other researchers (Korsching and Gildner 1986; Leistritz and Ekstrom 1986) in pointing to the increasingly bimodal distribution of farm production units. In 1987, farms with gross sales of $500,000 or more accounted for 25 percent of the total sales of all farms in the North Central region but represented only 1 percent of the farm units. Farms with sales of $100,000 or more accounted for 70 percent of total sales but represented only 17 percent of the region's farms and ranches. The farm crisis appears to have furthered the trend whereby increasing proportions of agricultural resources and production are coming under the control of a smaller number of production units.

At the same time, the findings show that substantial numbers of farm households in all 12 states in the region had managed to continue farming despite the difficult conditions of the 1980s, largely by obtaining off-farm employment to supplement their farm income. The ability of many of these households to continue farming may depend on the continued availability of such employment. With 60 percent of the survey households engaged in off-farm work, it is clear that many farm families have a vital stake in rural economic development.

Community services and infrastructure also concern both farm and non-farm residents of rural areas. The changes in availability and quality of services, as reflected in the survey responses, are due to changes in a number of factors including population size and composition, consumer tastes and shopping patterns (Ayres, Leistritz, and Stone 1992), the technology of service provision (e.g., medical care), and government regulations (e.g., banking). Whatever the causes of past changes, however, policy makers must determine how reasonable, affordable access to services can be provided to residents of rural areas.

The farm crisis of the 1980s came as a surprise to policy makers largely because they lacked information about the financial status of the nation's farmers. Subsequent attempts to respond to the economic restructuring were hampered by lack of information on which to base effective policies (Murdock and Leistritz 1988). As rural communities strive for economic development and diversification, additional types of information will be needed for success. For example, as farm operators attempt to enhance the returns from their resources (and thus increase their family income), information about the costs, returns, resource requirements, and risks associated with alternative enterprises will be important. Similarly, the potential of low-input, sustainable agricultural practices to enhance farm income and/or reduce risk will be important for many producers; the potential impact of the widespread adoption of such practices will interest the agribusiness sector. (When asked to name the topics about which they felt a need for education and training, the operators surveyed most frequently selected low-input farming methods, marketing skills, and effective use of new technologies.)

As an alternative to adding new enterprises to the farm operation, some farm families may wish to pursue a home-based business or other non-farm business venture. Information on the potentials of different types of businesses in rural areas, as well as factors to consider in embarking on such a venture, would be most useful. Also, policies and programs that support women's entrance into fuller partnership in the farm enterprise would assist farm women and farming.

The need to cushion income variability and volatility for farm households whose incomes depend mostly on farming is a policy issue. In Chapter 8, Meyer discussed stress among farmers. Stress among farm women, whose labor traditionally has underwritten the costs of direct food production in times of income variation, has been noted in other studies as well (Lobao and Meyer 1991). Even though the median household incomes and assets of Midwestern farmers were objectively adequate, income fluctuations made farm finances difficult for them. Swanson and Skees (1991) discuss strategies for moderate-sized commercial farms that reduce income volatility without bidding program benefits into land costs. Such strategies, which address the lack of predictability in farming sales, would aid the farm enterprise.

As stated previously, we believe that rural areas need more than merely farm programs; they need comprehensive rural development programs. These programs need to take account of the continued uneven development of the farm sector. Numerous authors have discussed the issues associated with designing development strategies for rural areas (see, for example, Pulver and Dodson 1992). In developing effective economic development strategies, however, a central element will be a wide variety of information for decision makers regarding community resources and development alternatives (Leistritz and Hamm 1994). Community leaders will need information about the area's currently available labor force, their education and job-related skills, and their willingness to accept employment under specified conditions (e.g., wage level, hours, type of work). Farm women, for instance, have problems in common with other women in the work force. They are concentrated in lower-paying, feminized occupations; they have difficulty in finding adequate child care; and farm and family responsibilities interact with employment opportunities and conditions (Lobao and Meyer forthcoming). Community development practitioners also must learn more about the economics of retail and service businesses located in small towns and rural areas: for instance, how can small-town retailers compete most effectively with large discount stores?

Economic development strategies and initiatives aimed at enhancing local employment opportunities would particularly benefit farm women because of their substantial and increasing involvement in off-farm work (Lobao and Meyer forthcoming). As noted previously, about one-half of all spouses in our survey had worked off the farm during the previous year. Further, off-farm work typically is more prevalent among younger farm women and among those in households that are just beginning farming careers. For example, a 1989 study of North Dakota households that had entered farming within the past five years revealed that 61 percent of the spouses had worked off the farm during the past year (as had 41 percent of the operators) (Leistritz et al. 1989).

Public service provision is yet another area of concern for community developers. What are the alternative means of providing specified services to residents in sparsely populated rural areas? What costs are associated with each alternative? This issue is of particular concern in non-metropolitan areas because residents of such areas typically have less access to services than do their metro counterparts. Such differences have been observed with respect to education (Swaim and Teixeira 1991) and health care (Hamm et al. 1993; U.S. Senate 1988) as well as for other types of services.

Education merits special attention. Skills and flexibility in the work force are a key (some would say the key) to competitiveness in the present-day economy, but rural areas and their workers typically have been at a disadvantage in this regard (Swaim and Teixeira 1991). Local governments in non-metropolitan areas typically have fewer resources for funding education than do their metro counterparts, and employers in non-metro areas have only limited capacities to upgrade workers' skills. Greater access to education and training could enhance farm residents' ability to participate in non-farm employment, and also could improve prospects for economic development in rural areas.

These are only a few of the questions that must be addressed if researchers are to contribute appreciably to developing effective policies for agriculture and for rural communities.

References

Ayres, Janet S., F. Larry Leistritz, and Kenneth E. Stone. 1992. *Revitalizing the Retail Trade Sector in Rural Communities: Lessons from Three Midwestern States*. RRD 162. Ames: Iowa State University, North Central Regional Center for Rural Development.

Hamm, Rita R., JoAnn M. Thompson, Janet K. Wanzek, and F. Larry Leistritz. 1993. *Medical Services in North Dakota*. Agricultural Economics Statistical Series No. 52. Fargo: North Dakota State University, Department of Agricultural Economics.

Korsching, Peter F. and Judith Gildner, eds. 1986. *Interdependencies of Agriculture and Rural Communities in the Twenty-First Century: The North Central Region*. Ames: Iowa State University, North Central Regional Center for Rural Development.

Leistritz, F. Larry and Brenda L. Ekstrom, eds. 1986. *Interdependencies of Agriculture and Rural Communities: An Annotated Bibliography*. New York: Garland.

Leistritz, F. Larry and Rita R. Hamm. 1994. *Rural Economic Development, 1975-1993: An Annotated Bibliography*. Westport, CT: Greenwood.

Leistritz, F. Larry, Brenda L. Ekstrom, Janet Wanzek, and Timothy L. Mortensen. 1989. *Beginning Farmers in North Dakota*. Agricultural Economics Report No. 249. Fargo: North Dakota State University, Department of Agricultural Economics.

Leistritz, F. Larry and Steve H. Murdock. 1988. "Financial Characteristics of Farms and of Farm Financial Markets and Policies in the United States." Pp. 13-28 in *The Farm Financial Crisis: Socioeconomic Dimensions and Implications for Producers and Rural Areas*, edited by Steve H. Murdock and F. Larry Leistritz. Boulder: Westview.

Lobao, Linda M., and Katherine Meyer. 1991. "Consumption Patterns, Hardship, and Stress among Farm Households." Pp. 191-209 in *Research in Rural Sociology and Development*, Vol. 5. Greenwich, CT: JAI Press.

Lobao, Linda M. and Katherine Meyer. Forthcoming. "Restructuring the Midwestern Rural Economy: Consequences for Farm Men and Women." *Economic Development Quarterly*.

Murdock, Steve H. and F. L. Leistritz. 1988. "Policy Alternatives and Research Agenda." Pp. 169-184 in *The Farm Financial Crisis: Socioeconomic Dimensions and Implications for Producers and Rural Areas*, edited by Steve H. Murdock and F. Larry Leistritz. Boulder: Westview.

Pulver, Glen and David Dodson. 1992. *Designing Development Strategies in Small Towns*. Washington, DC: The Aspen Institute.

Swaim, Paul and Ruy A. Teixeira. 1991. "Education and Training Policy: Skill Upgrading Options for the Rural Work Force." In *Education and Rural Economic Development*. ERS Staff Report AGES9153. Washington, DC: U.S. Department of Agriculture, Economic Research Service.

Swanson, Louis E. and Jerry R. Skees. 1991. "Issues Facing Agricultural Policy." Pp. 60-75 in *Rural Policies for the 1990s*, edited by Cornelia B. Flora and James A. Christenson. Boulder: Westview.

U.S. Senate Special Committee on Aging. 1988. *The Rural Health Care Challenge*. Serial No. 100-N. Washington, DC: U.S. Government Printing Office.

12

Methodology

Paul Lasley

Documenting the socioeconomic consequences of the farm crisis from a regional perspective presented numerous methodological and substantive challenges. Although the principal investigators had conducted state-wide surveys within our respective states (Ohio, Iowa and North Dakota) on the impacts of the farm crisis and rural restructuring, we lacked convincing and comparable data to make the case that farm restructuring was occurring throughout the North Central region. One objective of the North Central Regional Research Project (NC-184) was to assess the magnitude and impacts of the farm crisis. Funding for the study was provided through the North Central Regional Center for Rural Development.

Selection of Collaborators

There is a tendency to view the North Central region as relatively homogeneous, ignoring the considerable diversity in its environmental resources, agronomic, and cultural practices. From the dairy region in Wisconsin and Minnesota, to the wheat belt in the Dakotas, the small farm agriculture in the Missouri Ozarks, to the traditional corn-hog farms in the corn belt, it was necessary to have a broad based survey instrument that would be suitable across this wide range of farming systems. To conduct a regional survey that would be sensitive to the diverse set of issues within each subregion or state, collaborators in each of the 12 North Central states were solicited. Each of the

Agriculture Experiment Station Directors in the region was contacted and asked to appoint someone from their faculty to serve as a collaborator on the project. While the four principal investigators and the nine collaborators were all social scientists, they represented a wide diversity of perspectives and expertise, requiring that before the project could move forward, considerable time would have to be invested in forming a research team. One of the first tasks was to achieve consensus among the investigators on the conceptual approach, substantive issues of focus, and appropriate methods. While several of the scientists knew each other to varying degrees, they had little prior formal collaboration. Before data collection could occur, it was necessary to engage in team building exercises so that individuals could find common ground to cooperate on the project. Cooperation among the research team was achieved through fostering a climate of communication and joint decision-making. Within the confines of the research proposal, there was some latitude about the procedures and methods. Through a series of face-to-face meetings, telephone conference calls, and circulating memos during the summer and fall of 1988, the research team began to emerge as individuals shared their interests, perspectives, and what they hoped to gain from the project, either for their own professional interests or what it might provide for their state. In order to gain the active participation in the survey by all the scientists, it was necessary that everyone be given an opportunity to address particular issues and concerns relevant to their discipline or state. At the same time it was important that the focus of the project not be lost by including so many issues that the central focus of the project was diluted or lost.

Disciplinary Contributions

Each discipline represented on the research team brought a unique intellectual tradition to bear upon understanding the farm crisis, widely differing views about farm restructuring, and alternative interpretations about the potential social implications. The synergism of interdisciplinary research that is possible by bringing together diverse interests and insights requires that considerable time be spent in team building exercises to establish trust and explore areas of common interest. The breadth of perspectives, while illuminating the complexity of studying farm restructuring, also required much interaction to occur before the set of individual scientists could function as a team. Agricultural economists tended to focus on farm financial status and production data. Home economists and sociologists tended to focus on familial issues surrounding the farm household and the impacts of financial stress on the household functioning. Rural sociologists tended to emphasize the non-economic aspects of rural life, especially the interplay between farm, household, and community and how these relationships were tempered by gender, class, and other demographic characteristics.

Designing Mail Questionnaires

Developing mail questionnaires to address the familial, farm, and personal adaptations to the farm crisis was a time consuming process. The proposal called for including both farm men and women in the study. It was recognized that often farm spouses have been ignored in understanding the impacts of financial distress. All members of the research team were encouraged to submit questions for possible inclusion. Because of the unique issues affecting both men and women and the excessive length if only one questionnaire was used, the research team decided that two instruments should be developed—one for the farm operator and one for the spouse. Recognizing that in some cases the wife or female farm partner may be the actual operator and the man or husband may be employed off the farm, we chose to simply refer to the spouse and operator. A reiterative process of drafting questionnaires and circulating them among the research team members, while a slow and somewhat cumbersome process, ensured that everyone had opportunity to suggest questions or ideas for the survey. More questions and ideas for the questionnaire were received than could be used in the final version. It was necessary to start winnowing down the questions once all the submitted questions were gathered. The survey instruments reflected concerns about economic, farm, and community situations, future plans and options, social psychological well-being, coping techniques and strategies, perceptions about personal life and community, and community and political involvement. The questionnaires provide a comprehensive picture of the economic, social, and socio-emotional aspects of families struggling to survive restructuring of the farm economy.

Sampling Frame

Undertaking a major survey of farm operators and spouses across 12 states and involving hundreds of respondents presents several sampling problems. Farming is one of the few occupations where sampling is done based upon employment status. For example, there is no comparable sampling frames for mechanics, beauty shop operators, or truck drivers. Sampling frames are difficult to develop and maintain for several reasons. First, since farming is generally self-employment and increasingly performed on a part-time basis, federal sources that ask for only one major occupation may often exclude those who combine farming with other activities. A major criticism of farm sampling sources is that only the occupation of the person designated as the operator is considered (generally the male) and often ignores the contribution of the spouse (generally the female). This bias has contributed to widespread recognition that the role of women has been ignored in assuming that the operator is male, when in fact, both the husband and wife often share in the management and operation of the farm. Second, estimates are that 3-5 percent of the operators leave farming each year which necessitates constant and systematic updates of any sampling frame. Individuals may remain on a sampling frame because they once operated a farm, but are now retired or may be renting out the land. Third, given that entering and exiting farming is a lengthy process and involves several years, it is difficult to decide exactly when a person either enters or exits farming. Fourth, still others, despite objective criteria, may not consider themselves farm operators even though they meet the Census of Agriculture definition of a farm. Fifth, most data collection efforts of farmers are at the state or county level. Regional samples are even more problematic because they span a number of states, are subject to greater potential unevenness with regard to their completeness, frequency of update, and definitional ambiguities.

This project was aided by an existing cooperative agreement between the Iowa Agricultural Statistics Service and Iowa State University. This cooperative agreement, that has been in place for several years, enables joint participation in data collection and analysis. The Agricultural Statistics Service generally provides the most comprehensive list of farms available and is designed to cover farms included in the Census of Agriculture. The Iowa Agricultural Statistics Service agreed to participate in the regional study and coordinate with its counterparts in each of the other states to have the sample of farm operators selected. Since each Agricultural Statistics Service operates somewhat independently, each of the collaborators made personal contacts with them to secure their involvement in the project. The Iowa Agricultural Statistics Service agreed to assist in the coordination of the data collection and processing. In addition, because each Agricultural Statistics Service is operated from state offices, the frequency of updating the farm list varies by state. The research grant provided funds to select 400 households per state. In some cases, however, collaborators had additional funds and negotiated with their Agricultural Statistics Service to secure larger samples.

Data Collection

In February 1989, the two questionnaires were completed, printed, and ready to mail—one for the farm operator and one for the spouse or partner. In an attached letter we requested that "the person who is responsible for the majority of the farm operations decisions complete the farm operator survey and his/her spouse or household partner complete the other questionnaire." Single operators were instructed to complete only the operator survey. Mail questionnaires were sent to 11,800 farms in the region. Of the 11,800 farms that were sent questionnaires, 3,673 operators and 3,246 spouses responded, representing 3,940 households. Due to the aforementioned issues regarding sampling frames of regional farm populations, it is possible to only estimate rather that give a precise response rate. First, the Agricultural Statistics Service of each state vary in ability to update their sampling frames and cull those not farming in a particular year. For example, an Ohio study conducted in 1987 using a list of farm operators provided by the Ohio Agricultural Statistics Service found that 24 percent of all individuals on the list were not eligible for inclusion because they were not farming that year (Belyea and Lobao 1990). Second, due to new confidentiality requirements of the Office of Management and Budget that governs USDA, researchers are not given mail lists by Agricultural Statistics Service, but rather must rely upon them to send out questionnaires, which complicates in-depth follow-up that might increase response rates. Finally to increase response rates, if funds had been available, it would have been desirable to have multiple contacts with respondents, since follow-up solicitation improves response rates (Dillman 1978). Budget constraints and the complexity of working through 12 state Agricultural Statistics Services allowed for only one follow-up contact—a reminder postcard sent to all potential respondents.

Given these constraints, we calculate a range for the response rate: if one assumes the ideal that 11,800 farms were eligible for inclusion in the survey, the 3,940 households represents a very conservative estimate of 33 percent. If one assumes (based upon the Ohio study and recent experiences of other state and national surveys) that about one-fourth would have been ineligible because of retirement, obsolete lists, and natural attrition from retirement, then only 8,950 operators would have been eligible, and thus the response would be 45 percent. Accepting the most conservative estimate of a response rate of 33 percent compares favorably with other large-scale farm surveys employing a single follow-up. A recent national survey conducted in 1991 as part of a southern regional research project (S-246) employing five contacts with respondents, achieved only a 36.1 percent response rate (Molnar 1992).

In order to assess whether the findings reflected any response bias, telephone interviews with a random sample of non-respondents in each state were conducted. Response patterns between the mail questionnaires and telephone interviews with non-respondents were compared to assess if a response bias was introduced in the data collection process. No significant differences between mail questionnaires and telephone interviews were found among such background variables as farm size, debt level, household demographic characteristics, and across selected attitudinal items. Finding no differences between the two samples (mail questionnaires and telephone interviews), the samples were merged and yielded a total sample of 4,087 operators and 3,630 spouses.

Data Processing

Questionnaires were returned by the respondents in postage-paid envelopes to the Agricultural Statistics Service in their state. The Iowa Agricultural Statistics Service keypunched and verified the data, and assembled the data set. Each questionnaire was coded with state and respondent identification numbers so that spouses or farming partners could be linked together.

Weighting the Sample

Because farms are not distributed equally across the 12 North Central states, the sample sizes for the states were unequal. Accordingly, we weighted each state sample to provide a representative sample for the entire region. Our weighting procedure was based on the 1987 state Census of Agriculture reports. Tables 12.1 and 12.2 (Column 1) provide the number of farms for each of the 12 North Central states and the total number of farms for the North Central region. Column 2 contains the proportion of farms in the regional total represented by each state. Thus, the Illinois Census of Agriculture reported 88,786 farms in 1987, 10.2 percent of the total for the region.

TABLE 12.1 Information for Weighting Regional Samples (Operators)

State	Number of Farms	Percentage in Region	Number of Responses	Unweighted Percentage of Samples	Weight Factor (%Reg/%Sam)	Weighted Sample
Illinois	88,786	10.2	350	8.6	1.186	10.2
Indiana	70,506	8.1	367	9.0	.900	8.1
Iowa	105,180	12.1	398	9.7	1.247	12.1
Kansas	68,579	7.9	432	10.6	.745	7.9
Michigan	51,172	5.9	287	7.0	.843	5.9
Minnesota	92,000	10.6	303	7.4	1.432	10.6
Missouri	106,000	12.2	192	4.7	2.596	12.2
Nebraska	60,502	7.0	230	5.6	1.250	7.0
North Dakota	35,289	4.1	298	7.3	.562	4.1
Ohio	79,277	9.1	388	9.5	.958	9.1
South Dakota	36,376	4.2	207	5.1	.823	5.1
Wisconsin	75,131	8.6	634	15.5	.555	8.6
TOTAL	868,798	100.0%	4,087	100.0%		100.0%

Includes both mail and telephone responses.

TABLE 12.2 Information for Weighting Regional Samples (Spouses)

State	Number of Farms	Percentage in Region	Number of Responses	Unweighted Percentage of Samples	Weight Factor (%Reg/%Sam)	Weighted Sample
Illinois	88,786	10.2	315	8.7	1.172	10.2
Indiana	70,506	8.1	320	8.8	.920	8.1
Iowa	105,180	12.1	351	9.7	1.247	12.1
Kansas	68,579	7.9	408	11.2	.705	7.9
Michigan	51,172	5.9	249	6.9	.855	5.9
Minnesota	92,000	10.6	280	7.7	1.377	10.6
Missouri	106,000	12.2	166	4.6	2.668	12.2
Nebraska	60,502	7.0	222	6.1	1.147	7.0
North Dakota	35,289	4.1	243	6.7	.612	4.1
Ohio	79,277	9.1	353	9.7	.938	9.1
South Dakota	36,376	4.2	182	5.0	.840	5.1
Wisconsin	75,131	8.6	541	14.9	.577	8.6
TOTAL	868,798	100.0%	3,630	100.0%		100.0%

Includes both mail and telephone responses.

TABLE 12.3 Comparison of Personal and Farm Characteristics of Regional and Subregional Samples to U.S. Census of Agriculture

	NORTH CENTRAL REGION		CORN BELT		PLAINS		LAKES	
	Sample 4,087-O 3,630-S	Estimated Average*	Sample 2,112-O 1,875-S	Estimated Average*	Sample 947-O 844-S	Estimated Average*	Sample 1,028-O 911-S	Estimated Average
Average Age of Operator	52	50	52	51	52	50	51	50
Average Age of Spouse	49	NA	50	NA	50	NA	48	NA
Average Years of Education (Operator)	12	NA	12	NA	12	NA	12	NA
Average Years of Education (Spouse)	13	NA	13	NA	13	NA	13	NA
Average Size of Farm (Acres)	550	486	396	263	1,113	946	342	245

* The 1987 Census of Agriculture state averages for farm size and operator age were used to calculate estimated region and subregion averages.

O = Operator
S = Spouse

Table 12.1, Column 3, displays the number of operator responses for each state and gives the cumulative total. Table 12.2, Column 3, provides the number of spouse responses. Column 4 is the proportion of responses that each state contributed to the total number of respondents. Thus, for Illinois, 350 respondents represented 8.6 percent of the total sample. In this example, Illinois is slightly underrepresented (8.6% of the sample, compared with 10.2% of the region's population).

To correct for under and over-representation, we computed a weighting factor by dividing the percentage in the region (Column 2) by the percentage of the sample (Column 4). In each table, we used this factor to weight the responses for each question in the survey. The resulting individual weightings coefficients for states are shown in Column 5. In the case of Illinois, the weighting factor 1.186 is used to inflate the sample to achieve 415 respondents, or 10.2 percent.

To determine whether the weighted sample was representative of the region, we compared the distribution and means for farm size and operator age of the sample with the 1987 Census of Agriculture (Table 12.3). The average age of operators in the region was 50; our sample (collected two years later) was 52. The average farm size in the region according to the Census was 486 acres compared to 497 acres in our sample. Few differences existed between the sample and the general farm population in the region as reported in the Census of Agriculture.

As expected from other farm studies, few women designate themselves as farm operators – 98 percent of the spouses are women and 97 percent of the operators are men. Because of space limitations, not all questionnaire items could be addressed to both the operators and the spouses. We acknowledge reporting by spouses and operators as equally valid and as providing fuller information than could be furnished by one person. In some cases we asked spouses and operators the same questions so that we could make comparisons on conditions such as stress and family well-being. At other times we wanted to focus on aspects of farm life relevant to one individual, as in the case of what farm tasks were performed by women. Finally, we found no compelling reason to ask both operator and spouse the demographic and production questions.

References

Belyea, Michael and Linda M. Lobao. 1990. "Psychosocial Consequences of Agricultural Transformation: the Farm Crisis and Depression." *Rural Sociology* 55 (1):58-74.

Dillman, Don A. 1978. *Mail and Telephone Surveys, The Total Design Method.* New York: John Wiley and Sons.

Joseph J. Molnar. Memorandum, dated July 14, 1992, to S-246 Regional Research Committee.

Appendix

Survey Questionnaires

NORTH CENTRAL REGIONAL CENTER FOR RURAL DEVELOPMENT
Iowa State University of Science and Technology
216 East Hall
Ames, Iowa 50011
515-294-8322

Dear Farm Family:

There is considerable national interest in the social and economic well-being of farm families. While there is general agreement that the last decade has brought many changes to the farm industry, there is little information about how farm families have adjusted to these changes. We are conducting a 12 state study on the well-being of farm families, funded by the North Central Regional Center for Rural Development. A sample of farm families in each of the north central states is being asked to complete identical questionnaires. The states participating in the study are Illinois, Indiana, Iowa, Kansas, Michigan, Minnesota, Missouri, Nebraska, Ohio, North Dakota, South Dakota, and Wisconsin.

There are two questionnaires in this packet—one to be completed by the farm operator and another to be completed by the spouse or household partner. We would like the person who is responsible for the majority of the farm operation decisions to complete the farm operator survey, and his/her spouse or household partner to complete the other questionnaire. You will note each questionnaire is numbered to enable us to match couples' responses and to ensure an adequate follow-up. If you are single, simply complete your questionnaire and indicate your marital status on the appropriate question.

You have been randomly selected from a list of all farm families in your state. Therefore, it is important that we hear from everyone. Let me emphasize that your responses will be kept in strict confidence and information about individual respondents will <u>never be given out to anyone for any reason</u>. In addition, we have provided separate envelopes for each person to protect individual confidentiality.

Please complete and return your questionnaire as soon as possible. The information you provide will be used by researchers, Extension staff, and others who deliver services and programs to farm families. On behalf of the research team, we sincerely hope you will participate in this survey, and we thank you for your help.

Sincerely,

Pete Korsching
Director

Enclosures

OPERATOR'S QUESTIONNAIRE

DIRECTIONS
Please circle the appropriate number to answer each question. Your first impression is the answer desired on the following questions.

1. How have the following services, facilities, and economic conditions changed in your community over the past five years? Would you say they have generally "improved," "remained the same," or "gotten worse?"

		Improved	Remained the Same	Gotten Worse	Uncertain	Not Available
a.	Quality of schools	1	2	3	4	5
b.	Job opportunities	1	2	3	4	5
c.	Health care services	1	2	3	4	5
d.	Child care facilities	1	2	3	4	5
e.	Shopping facilities	1	2	3	4	5
f.	Police and fire protection	1	2	3	4	5
g.	Adult education opportunities	1	2	3	4	5
h.	Banking services	1	2	3	4	5
i.	Opportunities for entertainment and recreation	1	2	3	4	5
j.	The current financial condition of farmers has	1	2	3	4	5
k.	The current financial condition of agribusiness firms in your area has	1	2	3	4	5
l.	The current financial condition of lenders in your area has	1	2	3	4	5
m.	Your farm's financial condition has	1	2	3	4	5

2. Would you recommend farming to your children or another relative?

Yes	No	Not Sure
1	2	3

3. How often do you work with other people in your community to solve local problems?

At Least Once A Week	At Least Once A Month	At Least Once or Twice a Year	Not At All
1	2	3	4

4. Suppose that for some reason you had to move away from here. How sorry would you be to leave?

Very Sorry	Somewhat Sorry	A Little Sorry	Not At All Sorry
1	2	3	4

5. Please circle the response that comes closest to your opinion about the quality of life in your community.

	Become Much Better	Become Somewhat Better	Remained the Same	Become Somewhat Worse	Become Much Worse
a. During the past five years, **your** family finances have	1	2	3	4	5
b. During the past five years, the quality of life for **your** family has	1	2	3	4	5
c. In the next five years, the overall economic condition of farmers will	1	2	3	4	5
d. Considering your farm's overall financial situation, the likelihood that you will continue to farm for at least the next five years has	1	2	3	4	5
e. Compared to farmers in your area, your financial situation has	1	2	3	4	5
f. All things considered, your satisfaction with farming has	1	2	3	4	5
g. "Neighboring" over the past five years has	1	2	3	4	5
h. Neighbors helping each other over the past five years has	1	2	3	4	5
i. Do you believe the things you have in common with people in your community have	1	2	3	4	5

6. Has your family made any of the following adjustments **because of financial need** in the past five years?

	Yes	No
a. Used savings to meet living expenses	1	2
b. Sold possessions or cashed in insurance	1	2
c. Purchased more items on credit	1	2
d. Postponed major household purchase(s)	1	2
e. Let life insurance lapse	1	2
f. Cut back on charitable contributions	1	2
g. Changed food shopping or eating habits to save money	1	2
h. Changed transportation patterns to save money	1	2
i. Reduced household utility use such as electricity, telephone	1	2
j. Postponed medical or dental care to save money	1	2
k. Cancelled or reduced medical insurance coverage	1	2
l. Borrowed money from relatives or friends	1	2
m. Fallen behind in paying bills	1	2
n. Decreased money saved for children's education	1	2
o. Children have postponed education	1	2
p. Spouse has taken off-farm employment	1	2
q. You have taken off-farm employment	1	2

	Greatly Increased	Somewhat Increased	Remained the Same	Somewhat Declined	Greatly Declined
7. a. Over the past five years, your personal level of stress has..	1	2	3	4	5
b. Your concern with your level of stress has	1	2	3	4	5
c. On a day-to-day basis, your stress has............	1	2	4	5	5

8. Many farmers believe that the risk in farming has increased during the past five years. In <u>Column A</u>, please indicate what changes you have made **to reduce risk** in your operation. In <u>Column B</u>, please indicate the changes you are **planning to make** in the next five years. (Please answer both columns A and B.)

	Column A Changes Made (1984-1988)		Column B Changes Planned (1989-1993)		
	Yes	No	Yes	No	Maybe
a. Diversified farm by adding new crops	1	2	1	2	3
b. Diversified farm by raising livestock	1	2	1	2	3
c. Paid closer attention to marketing.....................	1	2	1	2	3
d. Postponed major farm purchase(s)....................	1	2	1	2	3
e. Reduced long-term debt	1	2	1	2	3
f. Reduced short-term debt	1	2	1	2	3
g. Bought additional land	1	2	1	2	3
h. Sold some land	1	2	1	2	3
i. Rented fewer acres	1	2	1	2	3
j. Rented more acres	1	2	1	2	3
k. Started a new business (not farming)	1	2	1	2	3
l. Used the futures markets to hedge prices	1	2	1	2	3
m. Shared labor or machinery with neighbors	1	2	1	2	3
n. Transferred land back to lender	1	2	1	2	3
o. Sought training for new vocation.......................	1	2	1	2	3
p. Reduced expenditures for hired help	1	2	1	2	3
q. Kept more complete financial records	1	2	1	2	3
r. Changed from cash rent to crop share..............	1	2	1	2	3
s. Bought crop insurance	1	2	1	2	3
t. Reduced machinery inventory	1	2	1	2	3
u. Sought off-farm employment	1	2	1	2	3
v. Retire from farming	1	2	1	2	3
w. Quit farming	1	2	1	2	3

9. There are a number of government programs and laws designed to assist farmers. For each of the programs or laws listed below, please indicate whether you have participated in them over the last five years and how much help they provided.

	YES, I participated in this program and found it to be:			NO, I did not participate in this program because:			
	No Help	Some Help	A Lot of Help	Not Needed	Did Not Qualify	Not Available	Did Not Know About
a. Federal government commodity programs (e.g. Feed Grain, Dairy Support)	1	2	3	4	5	6	7
b. Conservation Reserve Program (CRP)	1	2	3	4	5	6	7
c. Loans from FmHA	1	2	3	4	5	6	7
d. Farmer/lend mediation services	1	2	3	4	5	6	7
e. 1988 Drought Assistance Act	1	2	3	4	5	6	7
f. Federal all-risk crop insurance	1	2	3	4	5	6	7
g. Chapter 11 bankruptcy (debt reorganization)	1	2	3	4	5	6	7
h. Chapter 12 (debt restructuring for farmers)	1	2	3	4	5	6	7
i. Vocational retraining/ education program for self or family member	1	2	3	4	5	6	7
j. Job Partnership Training Act or other off-farm job search assistance program	1	2	3	4	5	6	7
k. Mental health counseling for yourself or family member	1	2	3	4	5	6	7
l. Food Stamps	1	2	3	4	5	6	7
m. Fuel Assistance	1	2	3	4	5	6	7
n. Unemployment benefits	1	2	3	4	5	6	7
o. Income assistance (e.g. AFDC, SSI)	1	2	3	4	5	6	7
p. Financial analysis or counseling by Extension Service	1	2	3	4	5	6	7

10. In order to continue farming in the next five years, I will need information/training on:

		Not Needed	Low Need	Moderate Need	High Need	Very High Need
a.	Marketing skills	1	2	3	4	5
b.	Diversification of farm operation by adopting new crops and livestock	1	2	3	4	5
c.	Available government assistance	1	2	3	4	5
d.	Bookkeeping and financial systems	1	2	3	4	5
e.	Using appropriate conservation techniques	1	2	3	4	5
f.	Using new technologies as they become available	1	2	3	4	5
g.	Using new machines and chemical inputs to increase my production	1	2	3	4	5
h.	Reducing production costs through low-input farming methods	1	2	3	4	5
i.	Processing farm products on farm before selling	1	2	3	4	5

DEMOGRAPHIC AND FAMILY CHARACTERISTICS

Please tell us about your background.

11. What is your sex?   1. Male _____   2. Female _____

12. What is your marital status?
    1. Single, never married
    2. Married
    3. Separated/Divorced
    4. Widowed

13. a. In what county do you live? _____

    b. How long have you lived in this county? _____ years

14. a. In what year did you become a farm operator? _____ year

    b. What is most likely to happen to your farm when you retire?
       1. Farm will be sold outside the family.
       2. Farm will be operated by child/children.
       3. Farm will remain in family but operated by someone else.

Please tell us about your family. If you are married, be sure to answer the questions for your spouse.

15. What are the ages of: Farm Operator _____   Spouse _____

16. What is the highest level of education completed: Yourself _____ years   Spouse _____ years

17. How many persons in your household are:

    Under 5 years of age _____
    5 to 12 years of age _____
    13 to 19 years of age _____
    Over 19 years of age _____

		OPERATOR	SPOUSE
18. a.	In 1988, did you (or your spouse) work off the farm?	Yes (1) No (2)	Yes (1) No (2)
b.	How many miles did you commute one-way?	_____ miles	_____ miles
c.	How many years have you (or your spouse) worked off the farm?	_____ years	_____ years
d.	On an average, how many hours per week do you work at this job?	_____ hours	_____ hours
e.	How many weeks in 1988 did you work at this job?	_____ weeks	_____ weeks

f. What occupation was the most important off-farm job in 1988? (Please check the appropriate category.)

	OPERATOR	SPOUSE
Professional	_____	_____
Manager/Administrator	_____	_____
Sales/Clerical	_____	_____
Craftsman/Machine Operator	_____	_____
Transportation	_____	_____
Laborer/Service Worker	_____	_____
Nonfarm self-employed	_____	_____
Other (please specify)	_____	_____

FARM OPERATION

In order to measure the extent of changes in farm operations from 1984 to 1988, please answer the following questions for last year (1988) and your best recollection for 1984.

		1988	1984
19.	What percent of the labor used on your farm was provided by yourself and other family members?	_____ percent	_____ percent
20.	On the average, how many hours per week do you work on the farm?	_____ hours	_____ hours
21.	How many acres did you own?	_____ acres	_____ acres
22.	How many acres did you rent or least FROM others?	_____ acres	_____ acres
23.	How many acres of land did you rent or lease TO others?	_____ acres	_____ acres

24. Of the total acres you farmed in 1988 and 1984, how many were in:

		1988	1984
a.	Cropland (include set aside acres)	_____ acres	_____ acres
b.	Pasture and Hay	_____ acres	_____ acres
c.	Conservation Reserve Program acres	_____ acres	_____ acres
d.	Timber/forest	_____ acres	_____ acres

25. Please indicate the number of acres planted to the following crops in 1988 and 1984. If you did not raise these crops, please enter "0."

	1988	1984
Corn	_____ acres	_____ acres
Soybean	_____ acres	_____ acres
Wheat	_____ acres	_____ acres
Oats	_____ acres	_____ acres
Sorghum	_____ acres	_____ acres
Barley	_____ acres	_____ acres
Other	_____ acres	_____ acres

26. Please indicate the number of livestock/poultry produced in 1988 and 1984. (If you did not raise, please enter "0.")

	1988	1984
Beef Cows	_____ head	_____ head
Feeder Cattle	_____ head	_____ head
Dairy Cows	_____ head	_____ head
Sows	_____ head	_____ head
Ewes	_____ head	_____ head
Turkeys	_____ number	_____ number
Laying Hens	_____ number	_____ number

FINANCIAL CONDITIONS IN FARMING

There is a continuing debate on the financial health of farming. Some people argue that the problem is being exaggerated, while others claim the financial condition remains a very serious problem. To help us assess the financial conditions in farming, we'd like you to answer the following questions.

27. As of January 1, 1989, what was the estimated current market value of your farm assets (including land, buildings, machinery, and livestock)?

    1. Real Estate Assets $_____
    2. Non-Real Estate Assets $_____

28. The value of your January 1, 1989 total assets, compared to a year earlier has:

    1. Increased   ____
    2. Remained the Same ____
    3. Decreased   ____

29. As of January 1, 1989, what were your estimated total liabilities (including all debts for land, buildings, machinery, livestock, and unpaid bills)?

    1. Real Estate Assets $_____
    2. Non-Real Estate Assets $_____

30. Your debts as of January 1, 1989, compared to a year earlier have:

    1. Increased   ____
    2. Remained the Same ____
    3. Decreased   ____

31. What was the gross farm sales, including government farm payments, from your operation in 1988?

    1. Less than $10,000
    2. $10,000-$39,999
    3. $40,000-$99,999
    4. $100,000-$249,999
    5. $250,000-$499,999
    6. $500,000 or more

32. Which of the following categories comes closest to your net family income for 1988? (This includes off-farm employment, farming and non-farm income such as interest and Social Security.)

    1. A loss was realized
    2. $1-$9,999
    3. $10,000-$19,999
    4. $20,000-$29,999
    5. $30,000-$39,999
    6. $40,000-$49,999
    7. $50,000-$59,999
    8. $60,000-$69,999
    9. Over $70,000

33. What percent of your total family income for 1988 was derived from:

Farming (including government farm payments)	_____ percent
Off-farm employment (including self and spouse)	_____ percent
Other non-farm income (i.e. interest, Social Security)	_____ percent
TOTAL	100 percent

34. a. Do you anticipate you will apply for a new farm loan or add to existing farm loans for 1989?   1. Yes ___   2. No ___   3. Not Sure ___

  b. Do you anticipate any problem in securing adequate financing for farm operating expenses for Spring 1989?   1. Yes ___   2. No ___   3. Not Sure ___

  c. As of January 1, 1989, did you have any outstanding farm non-real estate loans?   1. Yes ___   2. No ___   3. Not Sure ___

  d. Are you current on your non-real estate loans?   1. Yes ___   2. No ___   3. Not Sure ___

  e. As of January 1, 1989, did you have any outstanding land (real estate) mortgages?   1. Yes ___   2. No ___   3. Not Sure ___

  f. Are you current on your land (real estate) mortgages?   1. Yes ___   2. No ___   3. Not Sure ___

Comments:

THANK YOU FOR YOUR ASSISTANCE

SPOUSE'S QUESTIONNAIRE

This is a survey to determine the activities of spouses in the day-to-day operation of their farms. This survey is being done in conjunction with the operator's questionnaire. There are a few questions identical on both; however, most are different. Your first impression is the answer desired on the following questions.

Please complete and return your questionnaire in the enclosed self-addressed return envelope. The information you provide will be kept strictly confidential and used only in State and Regional summaries. Please note there is no identification except a state code and sequential code to keep track of the questionnaire through the edit and summary programs and to match the operator and spouse questionnaires.

On behalf of the research team, we sincerely hope you will participate in this survey, and we thank you for your help.

Sincerely,

Peter F. Korsching
Director of the North Central Regional
Center for Rural Development

PK/jb
Enclosures

DIRECTIONS
Please circle the appropriate number to answer each question.

1. We are interested in the kinds of work you do on your farm. Please indicate whether you have performed the following duties and if your time devoted to these tasks has changed over the past five years.

		I Perform These Duties..				My Time on These Duties Has..		
		Always	Some-times	Never	Not Done	Increased	Stayed the Same	Decreased
a.	Field work	1	2	3	4	1	2	3
b.	Milked or cared for farm animals	1	2	3	4	1	2	3
c.	Run farm errands	1	2	3	4	1	2	3
d.	Purchased major farm supplies and equipment	1	2	3	4	1	2	3
e.	Marketed farm products through wholesale buyers or directly to consumers	1	2	3	4	1	2	3
f.	Bookkeeping and maintained records	1	2	3	4	1	2	3
g.	Done household tasks and/or child care	1	2	3	4	1	2	3
h.	Supervised the farm work of others	1	2	3	4	1	2	3
i.	Took care of a vegetable garden or animals for family consumption	1	2	3	4	1	2	3
j.	Worked at an off-farm job	1	2	3	4	1	2	3

2. For each of the following decisions, please indicate whether you usually make the decision, your spouse/someone else makes the decision, or you make the decision together with your spouse/someone else.

	Usually Myself	My Spouse or Someone Else	Myself and Spouse or Someone Else	Decision Has Never Come Up
a. Buy or sell land	1	2	3	4
b. Rent more or less land	1	2	3	4
c. Buy major household appliances	1	2	3	4
d. Buy major farm equipment	1	2	3	4
e. Produce a new crop or livestock	1	2	3	4
f. When to sell your agricultural products	1	2	3	4
g. Try a new agricultural practice	1	2	3	4

3. Please circle the response that comes closest to your opinion about the quality of life in your community.

	Become Much Better	Become Somewhat Better	Remained the Same	Become Somewhat Worse	Become Much Worse
a. During the past five years, **your** family finances have	1	2	3	4	5
b. During the past five years, the quality of life for **your** family has	1	2	3	4	5
c. In the next five years, the overall economic condition of farmers will	1	2	3	4	5
d. Considering your farm's over-all financial situation, the likelihood you will continue to farm at least the next five years has	1	2	3	4	5
e. Compared to farmers in your area, your financial situation has	1	2	3	4	5
f. All things considered, your satsifaction with farming has	1	2	3	4	5
g. "Neighboring" over the past five years has	1	2	3	4	5
h. Neighbors helping each other over the past five years has	1	2	3	4	5
i. Do you believe the things you have in common with people in your community has	1	2	3	4	5

4. a. Suppose that for some reason you had to move from here. How sorry would you be to leave?

Very Sorry	Somewhat Sorry	A Little Sorry	Not At All Sorry
1	2	3	4

b. How often do you work with other people in your community to solve local problems?

At Least Once A Week	At Least Once A Month	At Least Once or Twice A Year	Not At All
1	2	3	4

	Would Help	Probably Help	Probably Would Not Help	Would Not Help	Unsure
5. a. If a situation developed that would hurt your community, would residents in your community help in doing something about it	1	2	3	4	5
b. If a farmer in your community had his/her farm foreclosed and some of the community residents thought it was unfair and organized a protest, would the **others** help out?	1	2	3	4	5
c. If a situation developed that would hurt your community, would **you** help in doing something about it?	1	2	3	4	5
d. If a farmer in your community had his/her farm foreclosed and some of the community residents thought it was unfair and organized a protect, would **you** help out?	1	2	3	4	5

6. There are many pressures on farm families. How frequently do you experience the following pressures?

	Would Help	Probably Help	Probably Would Not Help	Would Not Help
a. Problems in balancing work and family duties	1	2	3	4
b. Conflict with spouse	1	2	3	4
c. Conflict with children	1	2	3	4
d. Adjusting to new government policies	1	2	3	4
e. Difficulty with child care arrangements	1	2	3	4
f. No farm help or loss of help when needed	1	2	3	4
g. Lacking control over weather and commodity prices	1	2	3	4
h. Insufficient support from spouse in farm or family duties	1	2	3	4
i. Indebtedness and debt-servicing problems	1	2	3	4

7. During the past twelve months, how often did your family **not** have enough money to afford the kind of:

	Very Often	Fairly Often	Not Very Often	Never
a. Food your household should have	1	2	3	4
b. Clothes your household should have	1	2	3	4
c. Medical care your household should have	1	2	3	4

8. Below is a list of the ways you might have felt or behaved. How often have you felt this way during the past week?

		Rarely	Sometimes	Occasionally	All the Time
a.	I felt hopeful about the future	1	2	3	4
b.	My sleep was restless	1	2	3	4
c.	I was happy	1	2	3	4
d.	I felt lonely	1	2	3	4
e.	I was bothered by things that usually don't bother me	1	2	3	4
f.	I had trouble keeping my mind on what I was doing	1	2	3	4
g.	I felt depressed	1	2	3	4
h.	I felt that everything I did was an effort	1	2	3	4
i.	I enjoyed life	1	2	3	4
j.	I felt sad	1	2	3	4

9. There are many ways of coping with serious farm problems such as drought and low prices. Listed below are some of these coping strategies. How often do you use any of them?

		Use A Great Deal	Use Quite A Bit	Use Somewhat	Never Use
a.	Participate in church activities	1	2	3	4
b.	Become more involved in activities outside the farm	1	2	3	4
c.	Notice people who have more difficulties in life than I do	1	2	3	4
d.	Tell myself that success in farming is not the only important thing in my life	1	2	3	4
e.	Remind myself that for everything bad about farming, there is also something good	1	2	3	4
f.	Put up with a lot as long as I make a living from farming	1	2	3	4
g.	Go on as if nothing is happening	1	2	3	4
h.	Make a plan of action and follow it	1	2	3	4
i.	Try to make myself feel better by eating, drinking, smoking, using medication, etc.	1	2	3	4
j.	Refuse to think about it	1	2	3	4
k.	Keep problems secret from others	1	2	3	4
l.	Seek support from friends and/or relatives	1	2	3	4
m.	Seek spiritual support from minister, priest, or other	1	2	3	4
n.	Talk to a family counselor or other mental health professional	1	2	3	4
o.	Don't expect to get much income from farming	1	2	3	4
p.	Try to keep my feelings to myself	1	2	3	4
q.	Talk to someone who can do something concrete about the problem	1	2	3	4
r.	Wish that the situation would go away or somehow be over with	1	2	3	4

	Greatly Increased	Somewhat Increased	Remained the Same	Somewhat Declined	Greatly Declined
10. a. Over the past five years, your personal level of stress has	1	2	3	4	5
b. Your concern with your level of stress has	1	2	3	4	5
c. On a day-to-day basis, your stress has	1	2	3	4	5

11. We would like to ask some questions about your social activities:

	None	1 - 2	3 - 5	6 - 9	10 or More
a. How many clubs, organizations, and other groups (such as bowling teams, church groups, PTA) do you belong to	1	2	3	4	5
b. About how many people do you know from whom you can expect real help in times of trouble	1	2	3	4	5
c. About how many relatives to you have that you feel close to	1	2	3	4	5
d. About how many close friends do you have — people you can talk to about personal problems	1	2	3	4	5

12. On average, how often have you seen your close friends or relatives in the past month?

More Than Once A Week	At Least Every Week	At Least Once or Twice A Month	Not At All
1	2	3	4

13. There are a number of farm or local organizations. Please indicate **both you and your spouse's** membership activity in the following:

	Yourself Member	Yourself Former Member	Yourself Never Member	Spouse Member	Spouse Former Member	Spouse Never Member
a. Any farm organization, such as National Farmers Organizations, Grange, Farm Bureau, National Farmers Union, Young Farmers and Farm Wives	1	2	3	1	2	3
b. Any women's branches of general farm organizations, such as Farm Bureau Women	1	2	3	1	2	3
c. Any commodity producers' associations, such as the American Dairy Association or National Wheat Producers Association	1	2	3	1	2	3
d. Any women's branches of commodity organizations, such as the CattleWomen or the Wheathearts	1	2	3	1	2	3
e. Women's farm organizations, such as Women for Agriculture, American Agri-Women, or Women Involved in Farm Economics	1	2	3	1	2	3
f. Farm political action groups, such as a state Family Farm Movement or National Save the Family Farm Coalition	1	2	3	1	2	3
g. Local governing board, such as school board or town council	1	2	3	1	2	3
h. Marketing cooperative	1	2	3	1	2	3
i. Farm supply cooperative	1	2	3	1	2	3

		Yourself		Spouse	
		Yes	No	Yes	No
14.	In the past five years, have **you** or **your spouse**:				
a.	Attended public meetings about farm or other issues	1	2	1	2
b.	Talked to or written to government officials about public or farm issues	1	2	1	2
c.	Signed a petition	1	2	1	2
d.	Become more active in political groups	1	2	1	2
e.	Participated in a protest over farm foreclosures	1	2	1	2

15. Until you were sixteen, did you live mostly on a farm?   1. Yes ___   2. No ___

16. With which political party do you identify?   1. Democrat   2. Independent   3. Republican

17. Does your family have a religious preference?   1. No ___   2. Yes ___   If yes, please circle the denomination.

  1. Episcopalian, United Church of Christ, Presbyterian
  2. Methodist, Lutheran, Disciples of Christ, Christian, Central Christian, Disciples of Christ, First Christian, Northern Baptist, Reformed (such as Christian Reformed)
  3. Southern Baptist, Church of Christ, Evangelical/Fundamentalist, Nazarene, Pentecostal/Holiness, Assembly of God, Church of God, Adventist
  4. Catholic
  5. Jewish
  6. Church of Latter Day Saints
  7. Jehovah's Witness
  8. Christian Scientist
  9. Unitarian-Universalist
  10. Other

18. If you worked off the farm or were self-employed in 1988, how much net income did you earn?
  1. Less than $2,500
  2. $2,500-$4,999
  3. $5,000-$9,999
  4. $10,000-$19,999
  5. $20,000-$29,999
  6. $30,000-$39,999
  7. $40,000-$49,999
  8. $50,000-$59,999
  9. $60,000-$69,999
  10. Over $70,000

19. What percent of your family's net income for 1988 was derived from your off-farm work?   _____ %

20. On average, how many hours per week do you work off the farm?   _____ hours

21. Would you recommend farming to your children or another relative?   1. Yes   2. No   3. Not Sure

	Yes, Land and Assets	Yes, Land Only	Yes, Assets Only	No
22. Is your name on a deed or title to any of the family farm land or assets...	1	2	3	4

23. What is your sex?   1. Male   2. Female

24. In what county do you live?   _____

Comments:

_____
_____
_____
_____
_____
_____
_____
_____
_____
_____
_____
_____

THANK YOU FOR YOUR ASSISTANCE

Index

Adjustments/adaptations
    by communities, 19, 36, 139-141, 185-186, 210
    by farm households, 13-14, 18, 127-128, 130-131, 132(table), 133-134, 135-136(tables), 137, 138(table), 139-141, 182, 185, 193-194, 195-196(tables), 197-198, 203, 211
    of farm operations, 18, 46, 71-73, 76, 78-79, 85-86, 210-211
    financial, 36, 72, 75-76, 78-81, 85, 131, 132(table), 133-134, 135(table), 138(table), 139-141, 195-196(tables), 197-198, 203
Agriculture
    dependence on, 1, 3, 6-7, 15, 17, 30, 40
    economic environment, 1-11
    policy, 10-11. *See also* Government policy/programs
    productivity, 40-41
    restructuring, 1-12, 16-17, 21-22, 32, 35-36

Bankruptcy, 6, 12, 185

Capital purchases, 6
Commodity
    prices, 73, 147, 149-150, 150(table)
    producers' associations, 188, 190, 199, 200, 213, 214(table), 216
    production, 22, 46, 47(table), 48-49
    programs, 2, 8
    sales, 41(table), 42, 44
Community
    attachment, 127-129, 130, 141, 168, 180, 187, 193, 194, 195-196(tables), 197
    changes, 15, 17, 17(fig.), 19-20, 167-168, 172, 173-174(tables), 176, 177(table), 179(table), 180-181
    definition, 15, 21
    developers, 218
    participation/work, 187-188, 190-191, 192(table), 195-196(tables), 197, 215(table)
    services, 210, 216
    solidarity, 180, 186, 215(table)
    support, 183-184
Cooperative Extension Service, 89, 93, 103
Cooperatives, 188, 190, 213, 214(table)
    marketing, 189(table), 190-191, 213, 214(table)
    spousal participation, 214(table)
Coping, 125, 140, 145-149, 151-156, 210
Crop insurance, 74(table), 76, 77(table), 80(table), 81, 82-84(table), 92(table), 93, 96, 98(table), 100, 102(table)
Decision-making, 18, 110, 112, 117, 118(table), 119, 124-125, 190
Denial/escape, 153(table), 153-155
Depression, 14, 20, 128, 144-149, 151-156

Disparities, 208-210
Diversify
    livestock, 72, 74(table), 77(table), 78, 80(table), 82-84(table), 92(table), 93, 95, 95(table), 97(table), 100, 101(table), 105, 112, 117, 118(table), 211
    adding crops, 72, 74(table), 77(table), 78, 80(table), 83-84(table), 92(table), 93, 95(table), 95-96, 97(table), 100, 101(table), 105, 112, 117, 118(table), 211(table)
Division of labor, 13-14, 109-112, 124-125, 147, 198
Domestic labor, 122, 124-125

Economic base
    farming-dependent, 3-4, 171-176
    government-dependent, 171-176
    manufacturing-dependent, 171-176
    mining-dependent, 171-176
    service-dependent, 171-176
    trade-dependent, 171-176
Economic environment, 10, 29-49
    equity, 5, 8, 30
    exports, 4-5, 8, 30, 32(fig.), 33
    farm prices, 5, 109
    land values, 5, 9, 12, 29, 33, 33-34(figures), 53
    value of U.S. dollar, 30
Economic hardship, 18, 127-128, 131, 133-134, 137, 139, 141, 147-149, 156, 198, 201-202
Economic strategies, 124, 140, 214(table), 217-218
Economists and Agricultural Economists, 3, 11, 17-18, 20, 22, 207-208, 210, 222
Education
    effect on labor, 18-19, 110, 126, 218
    level of, 113, 116, 116(table), 117, 195-196(table), 197, 229(table)
    need for, 88, 94, 103, 105, 217
    operator, 116, 116(table), 117, 138(table)
    postpone child's, 132(table), 135(table), 140
    programs/funding, 103, 171, 219
Environmental movement, 10-11

Farm Credit System, 2, 6
Farm crisis
    adjustments to, 71-75, 78-79, 104-105, 124-125
    causes of, 4-7, 53-54
    effects of, 2, 6, 11-21, 40, 109-111, 127-129, 137-141
    policy-making, 2, 5-6, 8-11, 217
    perceptions of, 2-3, 7-11
    responses to, 183-185, 191-192, 201-202
Farm crisis impacts
    on agriculture, 4-9
    on business, 6, 9-10
    on communities, 6, 9-10, 13-15, 19-20, 185-186, 209-210
    on families/households, 7, 12, 14, 126-129, 137-139, 141, 185-186, 202, 208-209
    on farm operators/operations, 2, 4-9, 13-16, 20, 40, 54, 72-73, 78-79, 109, 185-186, 201
    on farm finances, 12-13, 17-18, 71, 78, 105, 139-141, 208
    on spouses, 14-15, 109, 127, 183
Farm financial characteristics
    assets, 5, 59, 60(table), 63, 71, 73, 93, 200, 211, 217: non-real estate, 59, 60(table); real estate, 60(table)
    balance sheet, 59, 60-62(table), 63
    debt, 61-62(table), 63-64, 67, 75, 100, 194, 199, 200-201, 208, 211: long-term, 74(table), 75, 77(table), 80(table), 85, 91-92; non-real estate, 61-62(table); real estate, 59, 61-62(table); servicing, 71, 73, 75, 80(table), 91-92, 100; short-term, 74(table), 75, 77(table), 80(table), 85, 91-92, 214(table), 215-216
    debt-asset ratio, 54, 62(table), 63, 64(fig.), 65, 65(table), 67(table), 68, 131, 132(table), 133-134, 135(table), 137, 138(table), 139,

193-194, 195-196(tables), 208, 212
economies of size, 76
financial need, 129-130, 132(table), 140
gross farm sales, 41(table), 43, 54-55, 56(table), 57, 58(fig.), 59, 62(table), 64(fig.), 65(table), 67-68, 80(table), 100, 101-102(table), 104(table), 138(table), 195-196(tables), 211-213, 214(table), 215-216
insolvency, 5, 63-64, 67
liabilities, 59, 61(table), 63
liquidation, 5-6, 75
liquidity, 53
net farm income, 53, 57, 111, 116, 169, 169(table)
net family income, 54, 56, 56(table), 57-59, 68, 81, 84(table), 112, 129, 133, 137, 138(table), 139, 195-196(tables), 199-200, 208-212
non-farm income, 55, 56(table), 57, 116
solvency, 13, 53, 59, 63

Farm input industry, 89

Farm organizations
American Agriculture Movement, 199
American Agri-Women, 189(table), 199
Farm Bureau, 188, 189(table), 199-200
Farmers Union, 188, 189(table), 199-200
membership in, 189(table), 190-191, 194, 195-196(tables), 197-203
National Family Farm Coalition, 189(table), 199
National Farmers Organization, 188, 189(table), 199-200
North American Farm Alliance, 189(table), 199
participation in, 183, 186-199, 189(table), 191, 193, 197-200
political action organizations, 189(table), 198-200
Prairiefire Rural Action, 199
spousal participation in, 184, 188, 189(table), 190, 197-200
Women for Agriculture, 189(table), 199
Women Involved in Farm Economics (WIFE), 189(table), 199

Farm spouse
age, 112, 115, 115(table), 196(table), 214(table), 229(table)
community involvement, 191, 192(table), 193, 196(table), 197, 215(table)
coping, 146-149, 152-156
depression, 147-149
education, 116, 116(table), 117, 138(table), 229(table)
farm work, 81, 112, 113(table), 114, 117, 120, 121(table), 122, 123(table), 124
in farm decisions, 18, 110, 112, 114, 117, 118(table), 119, 124
off-farm employment, 109-111, 114-117, 118(table), 119-120, 121(table), 122, 1213(table), 124, 132(table), 135(table), 209, 211, 218
organizational membership, 188, 189(table), 194, 196(table), 199, 214(table)
political activism, 16-17, 192, 192(table), 193, 196(table), 197-200
stress, 82-83(table), 147-148, 154-156, 215(table)
view of financial situation, 81, 133, 147-150
well-being, 129-131, 132(table), 134(table), 136(table)

Farming
commitment to, 128-129, 130, 141
full-time farm, 55
full-time operator, 19, 46, 55, 63, 65, 68, 124, 149, 201
part-time farm, 43, 46, 55, 185
part-time operator, 1, 19, 55, 63, 68, 114, 125

satisfaction with, 129-131,
132(table), 133, 134-136(tables),
139, 193, 195-196(tables)
Farming life cycle, 90
Financial record keeping, 72, 74-75, 77,
79-80, 81-82, 84-85, 91, 92(table),
94, 95(table), 96, 98-99(table), 100,
102(table), 103, 104(table),
113(table), 114, 120, 121(table),
122, 123(table), 211
Fordism/Fordist, 9, 13
Foreclosure, 6, 12, 16, 53, 143, 183,
185, 192, 197
Futures market to hedge prices, 73,
74(table), 77(table), 79, 80(table),
83-84(table), 92(table), 98(table),
100, 102(table)

Gender
differences, 146, 149, 191
division of labor, 13, 113
roles, 12, 14-15, 18
Government policy/programs, 2, 5-6, 8-
11, 15, 39, 48, 105, 147, 149,
150(table), 199, 217, 219

Hired labor, 76, 85, 209, 212

Increased activity, 153(table), 153-155

Labor allocation, 109-110
Labor market, 38, 88, 181
non-farm, 202
urban, 43, 46
Land, 92(table), 93, 95-96, 97(table),
100, 101(table), 105
buy, 92(table), 95, 96, 97(table),
100, 101(table), 118(table)
cash rent, 74(table), 77(table),
80(table), 83-84(tables), 92(table),
98(table), 102(table)
land values, 5, 30, 32-33, 33-
34(tables), 53
rent, 72, 74(table), 77(table), 78-79,
80(table), 81, 83-84(tables), 85,
92(table), 95-96, 97(table), 100,
101(table), 117, 118(table), 119

sell, 6, 92(table), 93, 96, 97(table),
101(table), 112, 118(table)
transfer land to lender, 6, 74(table),
77(table), 80(table), 81, 83-
84(table), 92(table), 93, 97(table),
101(table)
Loans, 66, 66(table), 67
payment/repayment, 5-6, 66, 139
non-real estate, 66, 66(table)
real estate, 66, 66(table)
Local employment opportunities, 170-
173, 174(table), 175-178, 179(table),
180, 209, 218

Marketing
attention to, 72-73, 74(table),
77(table), 79, 80(table), 81,
82(table), 84(table), 85, 89, 91,
92(table), 94, 96, 98(table),
102(table), 103, 210-211
farm products, 89, 113(table), 114,
120, 121(table), 123(table)
improving skills, 73, 88, 94,
95(table), 96, 99(table), 100,
102(table), 103, 104(table), 105
Media, 5, 16, 140, 143-144, 183, 201
Metropolitan counties, 36-39, 171-172,
173(table), 175-176, 177(table), 178,
179(table), 180-181, 209-210
Midwest states, 35-38, 37(table), 40, 44,
46, 48, 143, 145, 207, 210, 217

Non-farm agricultural industry, 89
agricultural lenders, 5-6, 89, 170
farm business management
associations, 89, 100, 103
Non-metropolitan counties, 36-37,
37(table), 38-39, 172, 209, 219
North Central States/Region, 33-35, 40,
43-46, 47(table), 48, 54, 55(fig.),
58(table), 64, 68, 71-73, 74(table),
76, 77(table), 80(table), 85, 90-91,
92(table), 94, 95(table), 168-169,
191, 207, 210, 221, 226, 227-
229(tables)
corn belt, 35, 40-41, 46, 54-58, 60-
68, 74, 76, 90-91, 92(table), 94

-96, 95(table), 97-99(table), 101-102(table), 104(table), 105, 168, 180-181, 190, 229(table)
lakes, 41, 54-63, 65-68, 74, 76, 90, 92(table), 94-96, 95(table), 97-99, 101-102(table), 104(table), 105, 168, 181, 190, 229(table)
plains, 41, 46-47, 54-68, 74, 76, 78, 90, 92(table), 94-96, 95(table), 97-99(table), 101-102(table), 104(table), 105, 170-171, 181, 190, 229(table)
North Central Regional Farm Survey, 21, 53, 72, 79, 87, 128, 146, 148, 154, 167, 184, 188
organization of, 89-91
sample area, 54, 55(fig.)
sample size, 54

Off-farm employment, 13, 18, 43, 45-46, 54-59, 56(table), 74(table), 76-81, 77(table), 80(table), 82-84(table), 85, 88, 92(table), 93, 95, 97(table), 100, 101(table), 105, 110-112, 114-125, 131, 132(table), 135(table), 184-185, 193-194, 195-196(tables), 184-185, 209, 211-212, 216, 218
Operator
age, 76, 77(table), 77-79, 90, 96, 97-99(tables), 100, 105, 112, 115, 115(table), 137, 138(table), 229(table)
community involvement, 191, 192(table), 193, 195(table), 197
membership in organization, 188, 194, 195(table), 214(table)
political activism, 192, 192(table), 193, 195(table), 197, 199-200

Political
action groups, 191, 198-203
activism, 16, 184, 187, 192, 192(table), 195-196(tables), 197, 201, 203
mobilization, 183, 187, 199-203
participation, 184, 186-181, 182(table), 193, 195-196(tables), 198, 201-202
protest, 16, 143, 183-184, 187-188, 192, 192(table), 196(table), 197-202
responses, 15-16, 187, 192, 200, 203, 208
Population, 37, 181
community, 176, 179(table)
decline, 3-4, 6-7, 15, 35, 181, 210
elderly, 12
farm, 37, 203
metro/urban, 36-37, 171, 176, 177(table), 178, 179(table)
Midwest, 36-38
migration, 36-37
rural/non-metro, 37, 177(table), 178, 179(table)
Postponed major farm purchase, 73, 74, 77, 80, 82, 84, 86, 132(table), 135(table), 211, 215
Price, 5, 12, 30, 31(table), 33, 109
Producers
agriculture, 19, 29, 40, 44, 88-89, 117, 118(table)
cattle/calves, 47(table), 48
corn, 46, 47(table), 48, 76
dairy, 47(table), 48
highly leveraged, 64, 211
hogs, 47(table), 48
soybean, 46, 47(table), 48, 76
wheat, 46, 47(table), 48, 76
Profitability, 53

Quality of life, 14-15, 127-129, 131, 132(table), 133, 134(table), 135-136, 136(table), 139-141, 185, 195-196(tables), 208, 213
Quit/retire from farming, 12-13, 45-46, 74(table), 77(table), 79, 80(table), 83-84(table), 88, 92(table), 93, 96-97, 97(table), 100, 102(table), 105, 139, 145-146, 156, 211

Reality construction, 154-155
Reduction of expenditures, 74(table), 76, 77(table), 80(table), 81, 82-84(tables), 85, 88, 92(table), 93,

98(table), 100, 102(table), 211, 214(table), 215-216
Relocation, 38-39
Restructuring
  economic, 4, 6, 36, 143, 185, 217
  farm, 1-4, 7, 11-12, 16-17, 143, 175, 183-185, 211, 221-222
  impacts of, 156, 167, 182, 191, 193, 201, 207
  industrial/manufacturing, 35, 175, 181
Risk reduction, 9, 71-73, 74(table), 75-76, 77(table), 78-79, 80(table), 81, 84(table), 85, 89, 92-93, 217
Rural restructuring, 139, 143, 145-146, 156, 175, 178, 181-182, 184-185, 191, 193, 201, 211, 221-222
Rural Sociologists and Sociologists, 3, 11, 16-20, 22-23, 128, 207-208, 210, 222-223

Satisfaction with farming, 129-133, 134(table), 135-136, 136(table), 139, 195-196(tables), 198-200, 208, 214(table)
Share labor or machinery, 74(table), 75, 76, 77(table), 79, 80(table), 81, 82-84(tables), 85, 92(table), 93, 96, 98(table), 102(table)
Size of farm, 7, 10, 19, 22, 41(table), 42, 44, 45(table), 67, 72, 75, 79, 80(table), 93, 100, 229(table), 230
  bimodal distribution, 216
Size of family, 110, 112, 115(table), 116, 124, 139
Sources of farm family income, 18, 39-40, 45-46, 48, 93, 209, 211-212

Start new business, 72, 74(table), 77(table), 78, 80(table), 81, 83-84(table), 85, 88, 92(table), 93, 97(table), 101(table), 105
Stress
  emotional, 14, 18, 79, 81, 82-83(table), 127-128, 144-151, 154-156, 215, 215(table), 217
  financial/economic, 12, 53-54, 71-73, 82-83(table), 124, 128, 131, 139, 147-149, 185-186, 223
Support seeking, 146-147, 150(table), 152-153, 153(table), 155
Survival strategies, 109-110, 185, 210-211

Training
  needs for, 87-90, 94, 95(table), 96, 99(table), 100, 103, 104(table), 105, 217
  for new vocation, 72, 74(table), 77(table), 78, 80(table), 81, 83-84(table), 87-90, 92(table), 93-94, 97(table), 100, 101(table), 105
  Vocational-technical school, 88, 89, 94

Unemployment, 181, 185

Well-being
  community, 19, 185, 200-201
  family, 14, 87-88, 127-135, 140-141, 185-186
  financial, 5, 129, 140-141, 185-186
  social-psychological, 14, 223